Linear Regression:
A Mathematical Introduction

*For Joan Gujarati, Diane Gujarati-Chesnut, Charles Chesnut,
and my grandchildren, "Tommy" and Laura Chesnut,
and for her behind-the-scenes help, Karen Low.*

Linear Regression

A Mathematical Introduction

Damodar N. Gujarati

*(Professor Emeritus of Economics,
U.S. Military Academy, West Point, NY)*

Los Angeles | London | New Delhi
Singapore | Washington DC | Melbourne

Quantitative Applications in the Social Sciences

A SAGE PUBLICATIONS SERIES

Quantitative Applications in the Social Sciences

A SAGE PUBLICATIONS SERIES

FOR INFORMATION:

SAGE Publications, Inc.
2455 Teller Road
Thousand Oaks, California 91320
E-mail: order@sagepub.com

SAGE Publications Ltd.
1 Oliver's Yard
55 City Road
London EC1Y 1SP
United Kingdom

SAGE Publications India Pvt. Ltd.
B 1/I 1 Mohan Cooperative Industrial Area
Mathura Road, New Delhi 110 044
India

SAGE Publications Asia-Pacific Pte. Ltd.
3 Church Street
#10-04 Samsung Hub
Singapore 049483

Acquisitions Editor: . Helen Salmon
Editorial Assistant: Megan O'Heffernan
Content Development Editor: Chelsea Neve
Production Editor: Kimaya Khashnobish
Copy Editor: QuADS Prepress Pvt. Ltd.
Typesetter: C&M Digitals (P) Ltd.
Proofreader: Scott Oney
Indexer: Hyde Park Publishing Services
Cover Designer: Candice Harman
Marketing Manager: Susannah Goldes

Printed in the United States of America

Library of Congress Cataloging-in-Publication Data

Names: Gujarati, Damodar N., author.

Title: Linear regression : a mathematical introduction / Damodar N. Gujarati.

Description: Los Angeles : Sage, [2019] | Includes bibliographical references and index.

Identifiers: LCCN 2018012989 | ISBN 9781544336572 (pbk. : alk. paper)

Subjects: LCSH: Regression analysis. | Matrices. | Sampling (Statistics)

Classification: LCC QA278.2 .G8445 2019 | DDC 519.5/36—dc23
LC record available at https://lccn.loc .gov/2018012989

This book is printed on acid-free paper.

18 19 20 21 22 10 9 8 7 6 5 4 3 2 1

CONTENTS

LIST OF FIGURES

SERIES EDITOR'S INTRODUCTION

I am very pleased to introduce *Linear Regression: A Mathematical Introduction* by Damodar Gujarati, one of the best-known econometricians of our era. The volume is a succinct introduction to the mathematics and statistical theory that is the foundation for classical linear regression analysis. It could be a course supplement for an advanced undergraduate or early graduate class in linear models. Alternatively, instructors might find it useful as a main text for the increasingly popular bootcamps in mathematics and statistics offered at the beginning of PhD programs. Even those already trained in these methods but in need of a refresher will find it of value. The volume is a welcome addition to the QASS Series.

Linear Regression: A Mathematical Introduction is very well structured and proceeds logically and carefully from the simplest to more complex linear regression models. When I read the draft manuscript, it was easy to imagine Professor Gujarati standing in front of the class working through the proofs and derivations on the chalkboard (or screen, as the case may be). As he would in person, Professor Gujarati explains how he proceeds from one step to the next, with lots of hints and tips. Reflecting its pedagogical purpose, there are exercises at the end of each chapter. Truth in advertising: the volume is mathematical, and readers who already know matrix algebra will get the most out of it. For readers who are a little rusty, there is an appendix that provides a brief but extremely helpful review. The appendix is also an entry point for readers without much training in matrix algebra. (A more comprehensive introduction can be found in *Matrix Algebra*, QASS Volume 38, by Krishnan Namboodiri.)

Importantly, *Linear Regression: A Mathematical Introduction* contains a clear exposition of the assumptions underlying the linear regression model, the consequences of violating each one, and the modifications of ordinary least squares needed to estimate linear regression models when this occurs. The volume provides a particularly helpful discussion of endogenous regressors, a perennial problem in social science applications. Indeed, the volume is remarkably practical. Once the key terms are defined and the foundations are laid, Professor Gujarati has plenty of practical advice about how, for example, to diagnose heteroscedasticity problems, interpret a Wald test, or assess the strength of an instrumental variable. With the advent of sophisticated software for statistical analysis, it is possible to run regression analyses without knowing the assumptions on which estimation and inference are based. Such ignorance is harmful. This volume is an antidote.

Classical linear regression analysis is one of the workhorses of the social sciences. Look at any major core journal in a social science discipline and you will find plenty of applications. At least as important is that linear regression is the foundation on which many more advanced statistical techniques are built. Multilevel models, simultaneous equations, and structural equation models are just a few examples of techniques rooted in regression. A thorough understanding of classical regression is necessary for a thorough understanding of these and other statistical models. *Linear Regression: A Mathematical Introduction* helps build that foundation. Students of all ages and stages will benefit from this volume.

Barbara Entwisle
Series Editor

PREFACE

Regression analysis is one of the most widely and intensively used techniques of quantitative research in fields as diverse as economics, finance, accounting, marketing, politics, international relations, agriculture, medicine, and biology. In fact, it is used in any area of research where one is interested in studying the relationship between a variable of interest, called the response variable, and a set of predictor variables. Sir Francis Galton (1822–1911) used it in the study of heredity, particularly the height of grown-up children in relation to the height of their parents. He used the method of least squares, the workhorse of linear regression analysis, for this purpose. Since then, the methodology of regression analysis has been improved and developed in many ways. It is no exaggeration to say that regression analysis has become an integral part of almost all scientific disciplines.

There is a well-developed mathematical and statistical theory behind the commonly used regression techniques. Some of this theory is quite complicated. My primary objective in writing this "Green Book" is to explain this theory in a rigorous but approachable manner to a large group of students, researchers, and teachers in various disciplines. With the ready availability of user-friendly statistical packages, estimating regression models is not a daunting task. But blindly using these packages without understanding the underlying theory could be a fruitless task, and at times, it could lead to misleading conclusions and policy prescriptions.

In about 250 pages, I explain the basics of linear regression, that is, regression linear in the parameters. The book contains seven chapters and four appendices. The key features of this book are as follows.

Key Features of the Book

1. A concise discussion of the ordinary least squares (OLS) and both the small- and large-sample properties of the OLS estimators.

2. A concise discussion of the method of maximum likelihood (ML) and the small- and large-sample properties of ML estimators.

3. A concise discussion of the distribution theory and a discussion of the commonly used tests of significance.

4. A concise discussion of the generalized least squares (GLS).

5. A concise discussion of the method of instrumental variables (IV) in cases where the predictor variables are stochastic. The classical least-squares model assumes that the predictors are either independent or at least uncorrelated with the regression error term.

6. Four appendices on the basics of matrix algebra, essentials of large-sample theory, small- and large-sample properties of estimators, and important probability distributions.

7. Two extended examples that discuss the various methods discussed in the book.

The technical discussion of some of the topics is put in the appendices to the various chapters for the benefit of more advanced students.

For upper-level undergraduate students, this book will provide a solid introduction to the linear regression models. Graduate students, teachers, and researchers will find this book to be a quick reference for the major themes in linear regression analysis.

Two datasets to accompany the book are available on a website at: **study.sagepub.com/gujarati**.

ABOUT THE AUTHOR

Damodar Gujarati (M.B.A. and Ph.D., both from University of Chicago) is Professor Emeritus of economics at the United States Military Academy at West Point. Prior to that, he taught for 25 years at the Baruch College of the City University of New York (CUNY) and at the Graduate Center of CUNY. He is the author of Government and Business, (McGraw Hill, 1984), the best-selling textbook Basic Econometrics (5th edition, 2009, with co-author Dawn Porter), as well as Essentials of Econometrics (4th edition, 2009, also with co-author Dawn Porter), both published by McGraw-Hill, and also Econometrics by Example (2nd edition, 2014, Palgrave-Macmillan). His experience spans business, consulting, and academia.

ACKNOWLEDGMENTS

I am grateful to the following reviewers for their incisive comments and suggestions on the draft of the book which has materially improved topics included in the book.

Michael Lewis-Beck, The University of Iowa

Colin Lewis-Beck, Iowa State University

Wendy Martinek, Binghamton University–SUNY

Kirk Randazzo, University of South Carolina

Jay Verkuilen, The Graduate Center–CUNY David Weakliem, University of Connecticut

During the preparation of the book I received valuable advice from the following academics:

Frank Fabozzi, EDHEC Business School, Paris Michael Grossman, The Graduate Center–CUNY

Charles B. Moss, University of Florida

I also want to thank Professors Paul M. Kellstedt and Guy D. Whitten at the Department of Political Science, Texas A&M University, for their valuable guidance on the earlier drafts of the book.

I owe special thanks to Professor Randall Campbell of the College of Business at Mississippi State University for reading the manuscript and for checking the various mathematical and statistical formulas and correcting some of the errors.

For their behind-the-scenes help, I am thankful to Barbara Entwisle, Kenan Distinguished Professor of Sociology, University of North Carolina at Chapel Hill, and series editor of *Quantitative Applications in the Social Sciences* (QASS); to Helen Salmon, Senior Acquisitions Editor of Research Methods, Statistics and Evaluation at SAGE Publishing; to Megan OHeffernan, Editorial Assistant at SAGE Publishing; to Rajasree Ghosh at QuADS, for copyediting this book; and to Kimaya Khashnobish and Arnab Karmakar, Production Editors at SAGE publishing.

CHAPTER 1. THE LINEAR REGRESSION MODEL (LRM)

1.1 Introduction

Regression analysis is one of the most widely and intensively used techniques of quantitative research in fields as diverse as economics, finance, accounting, marketing, politics, international relations, agriculture, medicine, and biology. In fact, it is used in any area of research where one is interested in studying the relationship between a variable of interest, called the response variable, and a set of predictor variables. Sir Francis Galton (1822–1911) used it in the study of heredity, particularly the height of grown-up children in relation to the height of parents. The regression technique that was used to study this relationship was the method of least squares. Since then the methodology of regression has been improved and developed in many ways. It is no exaggeration to say that regression analysis has become an integral part of research in almost all scientific disciplines.

Just to give a few examples, linear regression has been used in analyzing stock market returns, in the analysis of production and cost functions, in analyzing fertility and mortality rates, in the analysis of investment functions, in the analysis of the relationship between sugar intake and diabetes, in the analysis of death rates in relation to several factors, in the analysis of women's participation rates in the labor force, in the analysis of housing starts in relation to several socioeconomic variables, in the analysis of the effects of cigarette smoking on various types of cancer, in the analysis of admissions to graduate schools, in the analysis of presidential popularity, in the analysis of mental health, in the analysis of credit ratings of corporations, and in the analysis of crime rates in various suburban areas. In short, regression analysis has been used in a variety of situations, where the interest is in studying the relationship between the variable of interest in relation to several factors appropriate for the particular subject.

My primary focus in this book is on **linear regression**, which is the workhorse of regression analysis in most applications. Linear regression is also the foundation of the **generalized linear models (GLM)**, which I do not discuss in this book, for that requires a separate book.

In this and the following six chapters, I discuss the nature of linear regression and its theoretical foundations. Although students in applied disciplines may be interested in seeing how regression analysis is used in practice, they might benefit from learning about the underlying theory.

We express a generic linear regression model (LRM) as follows:

$$Y_i = B_1 X_{1i} + B_2 X_{2i} + B_3 X_{3i} + \cdots + B_k X_{ki} + u_i \quad i = 1, 2, 3, \ldots, n \qquad (1.1)$$

In this model, Y is the dependent variable; alternative names are explained variable, predictand, **regressand**, response, endogenous variable, outcome, and controlled variable. In this book, we will use the term *regressand*, which is a rather neutral term.

In this model, $X_1, X_2, \ldots, X_k$ are called the explanatory variables. Alternative names are independent variable, predictor, **regressor**, stimulus, exogenous variable, covariate, and control variable. We will use the more neutral term *regressor*. Some of the regressors are quantitative, and some are qualitative, such as race, gender, religion, and nationality. Very often, such qualitative variables are represented by **dummy variables**, taking values of 1 or 0, with 1 indicating the presence of an attribute and 0 indicating its absence. Sometimes the dummy variables are multicategorical, as we will illustrate with a concrete example in Chapter 4.

The subscript i is the observation subscript. By convention, the subscript i is used if the data are **cross-sectional** and the subscript t is used if the data are **time series**. If the data involve both cross-section and time-series observations, we use the double subscript it, as in X_{kit}, meaning the ith and tth observations on the regressor X_k. The number of observations is denoted by n.

We call (1.1) an LRM, and the meaning of the term *linear* will be explained shortly.

Generally, the variable X_1 takes the value of 1 for each observation in the data. This is to allow for the intercept in the model. As a result, we can write (1.1) as

$$Y_i = B_1 + B_2 X_{2i} + B_3 X_{3i} + \cdots + B_k X_{ki} + u_i \qquad (1.2)$$

We call (1.2) a **k-variable regression model**. The actual value of k depends on the phenomenon of study. Initially, we assume that the values of the regressors are fixed. Given the fixed values of the regressors, we draw repeated samples of Y values. We call this setup the **fixed regressor case**. In Chapter 6, we will show what happens if the values of the X regressors are also drawn randomly. In this case, both the regressand and the regressors are drawn randomly. This is the case of the **stochastic regressor**. *Stochastic*, a Greek word, means relating to a process involving a randomly determined sequence of observations, each of which is considered as a sample of one element from a probability distribution.

Any model, however extensive, is not expected to explain the phenomenon of interest fully due to random or uncontrolled influences. To account for the unavoidable random variation, we add the random variable u to the model, which is called the **error term**. It represents all those factors that may affect the regressand but are not included in the model because their individual influence on the regressand is very small and collectively all these factors may cancel out each other. It is also called a **random error**. More accurately, it is called a **stochastic error term**. Note that the error term is also known as the **disturbance term**.

The coefficients, $B_1, B_2, \ldots, B_k$ are called the **regression parameters** or **coefficients**. In the LRM, it is assumed that the regression parameters are fixed numbers and not random, although we do not know their values.[1] Once we have a set of data, we will show how the values of the parameters are estimated. The coefficient B_1 is called the **intercept** and the coefficients B_2 through B_k are called the **partial regression coefficients**, for reasons to be discussed shortly. In practice, it is better to retain the intercept in the model, although there are situations where it may be suppressed, as will be discussed subsequently.

1.2 Meaning of "Linear" in Linear Regression

Before proceeding further, it is important to know the meaning of the term *linear*, for it can be interpreted in two different ways. The first and perhaps more "natural" meaning of linearity is that the dependent variable is a linear function of the explanatory variables as in (1.1) or (1.2). In this sense, the following regressions are not linear regressions:

$$Y_i = B_1 + B_2 X_1^2 + u_i \quad \text{or} \quad Y_i = B_1 + B_2 \frac{1}{X_i} + u_i$$

The second interpretation of linearity is that the dependent variable is a linear function of the parameters, as in (1.1) and (1.2); here both these functions are linear in the variables as well as the parameters. But now consider the following functions:

$$Y_i = B_1 + B_2^2 X_i + u_i \quad \text{or} \quad Y_i = \frac{1}{1 + e^{B_1 + B_2 X_i + u_i}}$$

---

[1]Followers of *Bayesian statistics* regard these parameters as random. In this book, we will not pursue Bayesian regression models. See, for example, Koop, G. (2003). *Bayesian econometrics*. Chichester, England: Wiley; Weakliem, D. L. (2016). *Hypothesis testing and model selection in the social sciences* (chapter 4). New York, NY: Guilford Press.

These functions are not linear functions of the parameters; they are nonlinear functions of one or more parameters.

For our purpose, from now on the term *linear regression* will mean a regression that is linear in the parameters; it may or may not be linear in the explanatory variables. This does not mean the two preceding models cannot be estimated, but they require different estimation techniques, which are beyond the scope of this book.

Written out fully, Equation (1.2) represents the following set of equations:

$$
\begin{aligned}
Y_1 &= B_1 + B_2 X_{21} + B_3 X_{31} + \cdots + B_k X_{k1} + u_1 \\
Y_2 &= B_1 + B_2 X_{22} + B_3 X_{32} + \cdots + B_k X_{k2} + u_2 \\
&\vdots \\
Y_n &= B_1 + B_2 X_{2n} + B_3 X_{3n} + \cdots + B_k X_{kn} + u_n
\end{aligned}
\tag{1.3}
$$

This system of equations can be written more compactly as

$$
\begin{bmatrix} Y_1 \\ Y_2 \\ Y_3 \\ \vdots \\ Y_n \end{bmatrix}
=
\begin{pmatrix} 1 & X_{21} X_{31} \cdots & X_{k1} \\ \vdots & \ddots & \vdots \\ 1 & X_{2n} X_{3n} \cdots & X_{kn} \end{pmatrix}
\begin{bmatrix} B_1 \\ B_2 \\ B_3 \\ \vdots \\ B_k \end{bmatrix}
+
\begin{bmatrix} u_1 \\ u_2 \\ \vdots \\ u_n \end{bmatrix}
\tag{1.4}
$$

$$
n \times 1 \qquad\qquad n \times k \qquad\qquad k \times 1 \qquad n \times 1
$$

which we write as

$$
y = XB + u \tag{1.5}
$$

where

$y = n \times 1$ column vector of observations on the dependent variable

$X = n \times k$ matrix of n observations on k regressors, which includes a regressor whose value is 1 for each observation. X is often called the **data matrix**.

$B = k \times 1$ column vector of the k unknown regression parameters

$u = n \times 1$ column vector of n errors or disturbances u_i

Note: We represent matrices by capital bold letters and vectors by lowercase bold letters.

Using the rules of matrix addition and multiplication, the reader should verify that the Equations (1.3) and (1.4) are equivalent. Equation (1.5) is the **matrix representation** of the LRM.[2]

Equation (1.5) is often called the **population regression model (PRM)**, for it purports to show the relationship between the regressand and the regressors in the population of interest of some phenomenon. The concept of population is general and refers to a well-defined entity (people, firms, cities, states, countries, etc.) that is the focus of a statistical or econometric analysis.

As Equation (1.5) shows, the PRM consists of two components: (1) a **deterministic component, XB**, and (2) a **random component, u**. As shown below, XB can be interpreted as the **conditional mean** of Y_i, $E(Y_i|X)$, conditional on the given X values. Given the values of the regressors, Equation (1.5) states that an individual Y value is equal to the mean value of the population of which it is a member, plus or minus a random term. Note that we are using X as shorthand for the X matrix, that is, values taken by the regressors included in the data matrix.

The regression coefficient B_1 gives the *mean or average* value of the regressand when the values of regressors X_2 through X_k are all set to zero for each observation. The regression coefficients, B_2 through B_k, as noted previously, are known as **partial regression coefficients**. Each partial regression coefficient gives *the rate of change in the mean value of the regressand* for a unit change in the value of the regressor associated with it, holding all other regressor values constant.

One rarely observes the whole population of interest. Usually, given the values of the regressors, we draw a random sample of Y values and estimate $E(Y_i|X)$. Based on this estimate, we try to infer the true value of $E(Y_i|X)$. This dual task of estimation and hypothesis testing is the essence of statistical inference in regression analysis and in the other areas of statistics.

1.3 Estimation of the LRM: An Algebraic Approach

If b is a $k \times 1$ vector of estimators of B, we can write the estimated model as

$$y = Xb + e \tag{1.6}$$

where e, called the vector of **residuals**, is the sample counterpart of u.

---

[2]For the benefit of readers who are not familiar with matrix algebra or whose knowledge of the subject has become a little rusty, Appendix A provides a summary of the major themes in matrix algebra.

Based on a random sample of y and fixed X, how do we estimate Equation (1.5)? That is, how do we estimate the parameters in B? A commonly used method to estimate the regression parameters is the method of **ordinary least squares (OLS)**.[3] To explain this method, we rewrite the sample counterpart of Equation (1.2) as follows:

$$Y_i = b_1 + b_2 X_{2i} + b_3 X_{3i} + \cdots + b_k X_{ki} + e_i \tag{1.7}$$

Equivalently, we can write Equation (1.7) as

$$e_i = Y_i - b_1 - b_2 X_{21} - b_3 X_{3i} - \cdots - b_k X_{ki} \tag{1.8}$$

Or, in matrix notation,

$$e = y - Xb \tag{1.9}$$

Equation (1.9) states that the residual is the difference between the actual Y values and the estimated Y values obtained from the regression model (1.2). Since we know the (sample) values of the regressand and the regressor, this error term depends on the values of the estimated parameters in the model. Therefore, our objective is to obtain values of the regression parameters that would make the sum of the residuals as small as possible, ideally zero. However, this is not a good criterion, for we can have some large positive residuals and some large negative residuals that could make the sum of residuals practically zero. Likewise, we could have small positive residuals and small negative residuals and the sum of residuals can also be zero.

To avoid the problem of the signs of the errors, instead of minimizing the sum of errors, in OLS we minimize the sum of squared residuals as follows:

$$\Sigma e_i^2 = \Sigma (Y_i - b_1 - b_2 X_i - b_3 X_{3i} - \cdots - b_k X_k)^2 \tag{1.10}$$

To get a visual picture of minimization of the squared residual sum of squares, consider Figure 1.1.

In Figure 1.1, we have hypothetical data on the dependent and explanatory variables, Y and X, respectively. Such a figure is known as a **scattergram**. The straight line shown in the figure is the regression line. Not all the

---

[3]OLS is a special case of the method of **generalized least squares**, which we will discuss in Chapter 5.

Figure 1.1 Hypothetical Scattergram

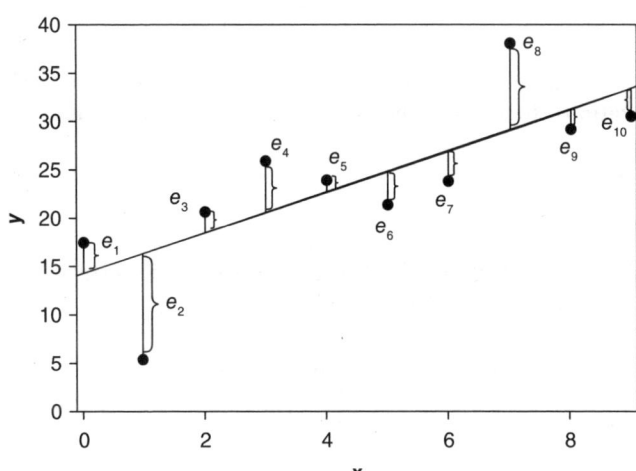

data points shown in the figure lie on this regression line. The vertical distance between the data points and the regression line are the e_is, the residuals. The regression line is drawn in such a way that the sum of the squared residuals is as small as possible. How this is done is shown in what follows.

The minimization of the sum of squared residuals can be handled by calculus techniques. Specifically, we differentiate Equation (1.10) with respect to the unknown bs, set the resulting equations to zero (the first-order condition of optimization), and rearranging, we obtain k equations in k unknowns, known as the **normal equations** of least squares (see Appendix 1A).

In matrix notation, we can write Equation (1.10) as

$$\Sigma e_i^2 = e'e = (y - Xb)'(y - Xb) \tag{1.11}$$

$$= y'y - 2bX'y + b'X'Xb \tag{1.12}$$

where use is made of the properties of the transpose of a matrix, namely, $(Xb)' = b'X'$, and since $b'X'y$ is a scalar (a real number), it is equal to its transpose $y'Xb$. Note that Σe_i^2 is called the **residual sum of squares (RSS)**.

To minimize $e'e$, we first differentiate Equation (1.12) with respect to b to obtain

$$\frac{\partial e'e}{\partial b} - 2X'y + 2X'Xb \qquad (1.13)$$

We now set these derivatives equal to zero (the first-order condition of optimization) to obtain the so-called **normal equations** of least squares:

$$X' - Xb = X'y \qquad (1.14)$$

Since X is $(n \times k)$, X' is $(k \times n)$. Therefore, $X'X$ is $(k \times k)$, a square matrix. If the inverse of $X'X$ exists, we can obtain

$$(X'X)^{-1}(X'X)b = (X'X)^{-1}X'y \qquad (1.15)$$

which reduces to

$$b = (X'X)^{-1}X'y \qquad (1.16)$$

This expression gives the estimators of the k unknown B coefficients. Notice that b is a **linear estimator**, that is, a linear function of the regressand y, which is obvious from this equation. Since y is random and the Xs are fixed, b is also random. B is, however, nonrandom. Note that Equation (1.16) gives the **point estimator** of each regression coefficient. That is, for a given sample, we obtain just one value, called the **estimate**, of each parameter included in the LRM. We can also obtain the **interval estimate** of a parameter, which gives a range that might include the true value of the parameter with certain probability. We will illustrate interval estimation in detail in Chapter 4.

Note: For the inverse of $X'X$ to exist, the matrix X must be of full (column) rank, k. This requires that the number of observations, n, must be greater than the number of parameters estimated, k in our case. In this case, $X'X$ is also of rank k.

To prove that b does minimize Equation (1.11), we can differentiate Equation (1.16) with respect to b, which yields

$$\frac{\partial e'e}{\partial b \partial b'} = 2X'X \qquad (1.17)$$

If the matrix X has full rank (i.e., rank k), the matrix of second-order (partial) derivatives given in Equation (1.17), called the **Hessian matrix**, is a **positive definite (PD) matrix**,[4] thus establishing that the b given in Equation (1.16) is indeed a minimum of Equation (1.12).

Note that $(X'X)$ is a symmetric square matrix of order $k \times k$. Written explicitly, it has the following form:

$$(X'X) = \begin{pmatrix} n & \sum X_{2i} & \sum X_{3i} & \cdots & \sum X_{ki} \\ \sum X_{2i} & \sum X_{2i}^2 & \sum X_{2i}X_{3i} & \cdots & \sum X_{2i}X_{ki} \\ \sum X_{3i} & \sum X_{3i}X_{2i} & \sum X_{3i}^2 & \cdots & \sum X_{3i}X_{ki} \\ \vdots & & & & \\ \sum X_{ki} & \sum X_{ki}X_{2i} & \sum X_{ki}X_{3i} & \cdots & \sum X_{ki}^2 \end{pmatrix} \qquad (1.18)$$

Notice these features of the $(X'X)$ matrix: (1) it is a symmetric matrix—the entries on either side of the main diagonal (running from upper left to lower right) are mirror images of one another; (2) the first row and the first column of this matrix give the sums of regressor values; note that n, the sample size, is the number of 1s added; and (3) the entries on the main diagonal give the sums of squares of the X variables and the off-diagonal entries give the sums of pairwise products of the regressors. As noted, the matrix $(X'X)$ is of the same rank as the matrix X.

We now express the estimated regression function as

$$\hat{y} = Xb \qquad (1.19)$$

which says y-hat or y-caret is the estimated (population) mean value of the dependent variable, given the values of the regressors. In other words, Xb is an estimator of XB.

Recall that

$$\begin{aligned} e &= y - Xb \\ &= y - X(X'X)^{-1}X'y, \quad \text{using Equation } (1.16) \\ &= My \end{aligned} \qquad (1.20)$$

---

[4] A symmetric matrix A is a positive definite matrix if $x'Ax > 0$ for all nonzero x. It is a positive semidefinite matrix if $x'Ax \geq 0$ for all x and there is at least one nonzero x for which $x'Ax = 0$. See Appendix A for further details.

where

$$M = I - X(X'X)^{-1}X' \tag{1.21}$$

where I is an $(n \times n)$ identity matrix, that is, a matrix with 1s on the main diagonal and zero elsewhere.

The M matrix has the following properties: It is square, symmetric ($M' = M$), idempotent (i.e., $M^2 = M$), singular, and of order n and rank $(n - k)$, and it has the property that $MX = 0$ (verify this).[5] As a result, we have

$$X'e = X'My = 0 \tag{1.22}$$

which shows that the regressors and the residuals are orthogonal. In words, each regressor and the associated residuals are independent. Note that M is *singular*, as its rank is $(n - k)$—the number of observations minus the number of regressors. The proof of this will be presented shortly.

The M matrix is also known as a **projection matrix**.

We can now write Equation (1.19) as

$$\hat{y} = Xb = Hy \tag{1.23}$$

where

$$H = X(X'X)^{-1}X' \tag{1.24}$$

H is called the "hat matrix" as it transforms y into y-hat ($\hat{y}$). Like M, H is also a **projection matrix**. Geometrically, it projects y (perpendicularly) onto $\hat{y}$ (y-hat).

It is easy to verify that

$$H' = H, \quad H^2 = H, \quad H + M = I, \quad \text{and} \quad HM = 0 \tag{1.25}$$

As a result,

$$y = Hy + My = \hat{y} + e = Xb + e \tag{1.26}$$

---

[5]For any idempotent matrix, its rank is equal to its trace. Here, $\text{tr}(X(X'X)^{-1}X') = \text{tr}(X'X(X'X)^{-1}) = \text{tr}(I_k) = k$. Hence $\text{tr}(M) = \text{tr}(I_n) - k = n - k$. That is why M is singular. Remember that a matrix whose determinant value is 0 is called a singular matrix.

Furthermore, because of Equations (1.20) and (1.22), it follows that

$$\hat{y}'e = 0 \tag{1.27}$$

That is, the vectors y-hat and e are orthogonal. In other words, the estimated Y values and the residuals are independent.

We can visualize this relation in Figure 1.2.

Figure 1.2 shows the vector y, the estimated y vector ($= Xb$), and the residual vector, e. As you can see, the residual vector (e) makes a right angle with the estimated y vector. This figure is an example of an **orthogonal projection**, or loosely speaking "dropping a perpendicular." Technically speaking, "a projection is a mapping that takes each point of E^n into a point in a subspace of E^n, while leaving all points in that subspace unchanged. Because of this, the subspace is called the **invariant subspace** of the projection."[6] The projection matrices discussed above perform this orthogonal projection.

In short, the matrix H applied to y gives the vector of fitted or estimated values of y and the matrix M applied to y gives the vector of least-squares residuals, e.

Note an interesting relationship between the projection matrices H and M:

$$H + M = I \tag{1.28}$$

Figure 1.2 Orthogonal Projection (Residuals and Fitted Values)

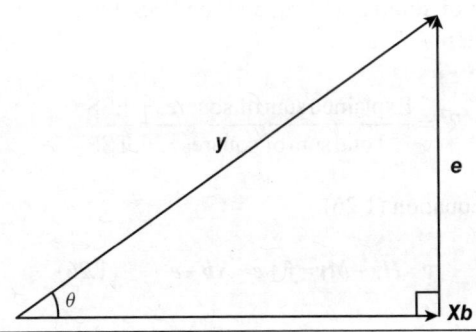

---

[6]Davidson, R., & MacKinnon, J. (2004). *Econometric theory and methods* (p. 57). New York, NY: Oxford University Press.

These two projection matrices are therefore **complementary projections** because

$$Hy + My = y \tag{1.29}$$

That is, H and M applied to y produce the original vector y, as can be readily verified.

Another interesting property of the two projection matrices is that their product produces the **null matrix 0**:

$$HM = 0 \tag{1.30}$$

In other words, the two projection matrices **annihilate** each other.[7] Every element of the null matrix is zero.

The matrices M and H play a vital role in analyzing LRMs, as the subsequent discussion will show.

1.4 Goodness of Fit of a Regression Model: The Coefficient of Determination (R^2)

Besides estimating the parameters of a regression model, we are often interested in finding how well the chosen regression model explains the variation in the regressand. For this purpose, we obtain a measure of goodness of fit, called the **coefficient of determination**, denoted by R^2. It measures the proportion or the percentage of variation in y explained by the regression model—that is, by the regressors included in the model.

To explain R^2, we develop the concepts of **total sum of squares (TSS)**, **explained sum of squares (ESS)**, and the RSS. Once these quantities are estimated, we define R^2 as

$$R^2 = \frac{\text{Explained sum of squares}}{\text{Total sum of squares}} = \frac{\text{ESS}}{\text{TSS}} \tag{1.31}$$

We start with Equation (1.26):

$$y = Hy + My = \hat{y} + e = Xb + e \tag{1.26}$$

---

[7]Davidson, R., & MacKinnon, J. (2004). *Econometric theory and methods* (p. 59). New York, NY: Oxford University Press.

Premultiplying both sides of this equation by y', we obtain

$$y'y = (Xb + e)'(Xb + e)$$
$$= b'X'Xb + e'(Xb + e)$$
$$= b'X'Xb + e'e, \quad \text{since } X'e = 0 \tag{1.32}$$

However, this is the raw sum of squares of the actual Y values, $y'y = \Sigma Y_i^2$. Verbally, we can write Equation (1.32) as

raw TSS = raw ESS + raw RSS

where TSS = total sum of squares; ESS = explained sum of squares, that is, that part of TSS explained by the regression model; and RSS = residual sum of squares, that is, that part of TSS not explained by the regression model.

The term *raw* means these sums of squares are not measured as deviations from their respective mean values.

If the regression model contains an intercept term, we would like to compute the sum of squares of Y values around its mean value, that is,

$$\Sigma(Y_i - \bar{Y})^2 = \Sigma Y_i^2 - n\bar{Y}^2 = y'y - n\bar{Y}^2 \tag{1.33}$$

Subtracting, $n\bar{Y}^2$ from both sides of Equation (1.32), we obtain

$$(y'y - n\bar{Y}^2) = (b'X'Xb - n\bar{Y}^2) + e'e \tag{1.34}$$
$$\text{TSS} = \text{ESS} + \text{RSS}$$

TSS is the mean-corrected total sum of squares of the regressand; ESS is the mean-corrected explained sum of squares, that is, that part of the TSS that is explained by the regressors in the model; and RSS, the residual sum of squares, is the remainder of TSS that is not explained by the regressors. If the estimated model fits the data well, we would expect ESS to explain a substantial variation in the regressand. The coefficient of determination is a measure of how well the fitted model explains the variation in the dependent variable.

Following Equation (1.31), we obtain

$$R^2 = \frac{b'X'Xb - n\bar{Y}^2}{y'y - n\bar{Y}^2} \tag{1.35}$$

R^2 is also defined as

$$R^2 = 1 - \frac{\text{RSS}}{\text{TSS}} = 1 - \frac{e'e}{y'y - n\bar{Y}^2} \tag{1.36}$$

The R^2 thus defined lies between 0 and 1.

It is a property of R^2 that it increases as additional regressors are added to the model. To compare regression models with the *same* regressand but differing number of regressors, researchers use a variant of R^2 known as the **adjusted R^2**, denoted by $\bar{R}^2$ (read as R-bar squared). It is defined as follows:

$$\bar{R}^2 = 1 - \frac{\Sigma e_i^2 / (n-k)}{\Sigma(Y_i - \bar{Y}^2)/(n-1)} \qquad (1.37)$$

This can be translated as follows:

$$\bar{R}^2 = 1 - (1 - R^2)\frac{n-1}{n-k} \qquad (1.38)$$

which not only shows the relationship between adjusted and unadjusted R^2 but also shows that for $k > 1$, $\bar{R}^2 < R^2$, which implies that as the number of regressors in the model increases, the adjusted R^2 increases less than the unadjusted R^2. The term *adjusted* means adjusted for the **degrees of freedom** (df)[8] for the sums of squares associated in Equation (1.34). So to speak, there is a penalty for adding more regressors to the model. Adjusted R^2 is often used to compare competing regression models. But it is important to note that to compare two or more regression models on the basis of $\bar{R}^2$, *the dependent variable must be the same*. Thus, if the dependent variable in one model is Y but log Y in another model, we cannot use $\bar{R}^2$ to compare the two models. This is because variation in Y and the variation in log Y are not the same. In the former, it is the absolute change, whereas in the latter it is the relative change.

A caution on the role of R^2 in regression analysis: The late Professor Arthur Goldberger has this comment about the R^2.

> From our perspective, R^2 has a very modest role in regression analysis, being a measure of goodness of fit of a sample LS [least square] linear regression in a body of data. Nothing in the CR [classical regression] model requires that R^2 be high. Hence a high R^2 is not evidence in favor of the model, and a low R^2 is not evidence against it. . . . In fact the most important thing about R^2 is that it is not important in the CR

---

[8]The degrees of freedom means the number of values free to vary when computing a statistic. For example, the mean value of five observations [5, 8, 10, 13, 14] is 10. To keep the mean value at 10, we can change the value(s) of only four number(s). So here we have only 4 degrees of freedom, although there are 5 observations.

model. The CR model is concerned with parameters in a population, not with goodness of fit in the sample.[9]

However, R^2 and adjusted R^2 have become a standard feature of most statistical packages. We discuss the classical LRM in Chapter 2.

1.5 R^2 for Regression Through the Origin

Occasionally, we come across a regression model without the constant, or intercept, term. Such a model is called **regression through the origin** or **zero-intercept model**. For this model, we use the raw sums of squares defined in Equation (1.32) and obtain what is called the **raw** R^2.

$$R^2_{\text{raw}} = \frac{\text{ESS}_{\text{raw}}}{\text{TSS}_{\text{raw}}} \tag{1.39}$$

The raw R^2 does not have the same properties as the traditional R^2. But we will have more to say about this measure in Chapter 2. *Suffice it to note here that unless there is a compelling reason to use the zero-intercept model, it is better to retain the intercept in practice.*

1.6 An Example: The Determination of Hourly Wages in the United States

Before we conclude this chapter, here is a concrete example of a linear regression. Based on a random sample of 1,289 workers from the Current Population Survey (CPS) for March 1995, we obtained the following regression based on the method of OLS.

$$W_i = -7.1833 - 3.0748FE_i - 1.5653NW_i + 1.0959UN_i + 1.3703ED_i + 0.1666EX_i$$

$$n = 1,289, R^2 = 0.3233 \tag{1.40}$$

where W=hourly wage in dollars, FE (gender), coded 1 for female, 0 for male, NW (race), coded 1 for nonwhite workers and 0 for white workers, UN (union status), coded 1 if in union job, 0 otherwise, ED (education) in years, EX (work experience) in years. In this regression, FE, NW, and

---

[9]Goldberger, A. S. (1991). *A course in econometrics* (p. 177). Cambridge, MA: Harvard University Press.

UN are **dummy variables** and *ED* and *EX* are quantitative variables. For discussion purposes, we call (1.40) a **wage regression**. Note that *EX*, the experience variable, is defined as age minus years of schooling minus 6; it is assumed that schooling starts at 6 years of age.

The details of regression (1.40) will be fully discussed in Chapter 4. But this is how we interpret this regression. The negative intercept value in this regression has no viable economic meaning, for literally interpreted, it suggests that if the values of all the regressors in this regression are held constant at zero, the average wage is a negative $7.18. In many regressions, the intercept value may not be meaningful.

The interpretation of the quantitative variables is straightforward. For example, the education coefficient suggests that if education increases by a year, the average wage goes up by about $1.37, ceteris paribus (holding the values of the other regressors in the model constant). Similarly, the coefficient of the experience variable suggests that the average wage goes up by about $0.16 per year of service experience, ceteris paribus. The positive value of these two coefficients makes economic sense.

The coefficients of the qualitative, or dummy, variables are interpreted differently. Here the comparison is between the variable that gets a value of 1 and the one that gets the value of 0. Thus, the negative coefficient of the gender variable suggests that female workers, on average, earn less than male workers by about $3. Values of other dummy coefficients should be interpreted similarly. Again, note that the negative coefficients of gender and race variables and the positive coefficient of the union variable are in accordance with labor market behavior.

The R^2 value of about 0.32 suggests that the predictors or regressors included in the wage regression explain about 32% of the variation in the average wage. This value might seem low, but we will have more to say about this in Chapter 4, and also note the caution sounded by Arthur Goldberger.

1.7 Summary

In this chapter, we introduced the *k*-variable LRM in its most general form. At the outset, we stated that by an LRM we mean a regression model that is linear in the parameters and not necessarily linear in the variables. We also discussed briefly the nature of the *stochastic* error term u_i and stated that it is an integral part of the LRM. We introduced the LRM both in the scalar form and in the matrix form. Once we go beyond the simple two- or three-variable regression models, we need to use matrix algebra to avoid lengthy algebraic equations and derivations.

In practice, we rarely observe the true population of interest. Invariably, we believe we have a random sample from the population of interest, and all the analysis is based on the sample data. For this purpose, we introduced the **sample regression model**, which we use to estimate the unknown population parameters. To take into account sampling variability, we introduced the sample error term, called the residual, e_i, as a proxy for the true error term, u_i.

To estimate the unknown population parameters, we used the method of OLS. In OLS, we minimize the RSS, Σe_i^2. The resulting estimators of the regression parameters are known as **OLS estimators**.

Two of the important properties of OLS are that the estimated Y values and the residuals are uncorrelated and that each regressor is uncorrelated with the respective residual, assuming the X are fixed or constant. As we will show in the next chapter, the mean value of the residuals $\bar{e}$ is zero.

It is interesting that so far we have not made any assumptions regarding the probabilistic properties of the error term, u_i. The only assumptions we have made are that the X matrix is nonstochastic and that the number of observations, n, is greater than the number of parameters estimated, which is another way of saying that the data matrix X is of full (column) rank.

The purely algebraic approach to estimate regression parameters is not adequate, for the objective of regression analysis is not only to estimate the parameters of the PRM, which is based on sample data, but also to draw inferences about the true values of the parameters. Estimates based on sample data are subject to variation from sample to sample.

To accomplish the twin objectives of estimation and inference, we need a framework. The **classical linear regression model (CLRM)** provides such a framework. We discuss the CLRM in Chapter 2.

Exercises

1.1 Which of the following models are linear in the parameters, or variables, or both. Which of these models are LRMs?

a. $Y_i = B_1 + B_2(1/X_i) + u_i$ (Reciprocal)

b. $Y_i = B_1 + B_2 \ln X_i + u_i$ (Semilogarithmic)

c. $\ln Y_i = B_1 + B_2 X_i + u_i$ (Inverse semilogarithmic)

d. $\ln Y_i = B_1 + B_2 \ln X_i + u_i$ (Double logarithmic)

e. $\ln Y_i = B_1 + B_2(1/X_i) + u_i$ (Logarithmic reciprocal)

Note: ln = natural log and u_i is the regression error term.

1.2 Are the following models LRMs? Why or why not?

a. $Y_i = e^{B_1 + B_2 X_i + u_i}$

b. $Y_i = \dfrac{1}{1 + e^{B_1 + B_2 X_i + u_i}}$

c. $\ln Y_i = B_1 + B_2 (1/X_i) + u_i$

d. $Y_i = B_1 + (0.75 - B_1)e^{-B_2(X_i - 2)} + u_i$

e. $Y_i = B_1 + B_2^3 X_i + u_i$

1.3 Consider the following regression model that has no explanatory variables.

$$Y_i = B_1 + u_i$$

a. Use OLS to estimate B_1.

b. How would you interpret B_1 in this model?

1.4 Consider the following simple two-variable, or bivariate, regression model.

$$Y_i = B_1 + B_2 X_i + u_i$$

a. Using OLS, obtain the estimators of B_1 and B_2.

b. How would you interpret the two regression coefficients?

1.5 Prove: $r_{Y\hat{Y}}^2 = R^2$, that is, the squared correlation coefficient between the actual Y and the estimated Y from a regression model is equal to the coefficient of determination.

1.6 Let B_{YX} be the slope coefficient in the regression of Y on X and B_{XY} the slope coefficient in the regression of X on Y. Show that $B_{YX} B_{XY} = r^2$.

1.7 Show that $\bar{X} = 1'x/n$. Similarly, show that $\bar{Y} = 1'y/n$. Show that $\sum_1^n X_i = 1'x$, where $1' = (1, 1, \ldots, 1)'$ and x is an n-element column vector.

1.8 Prove the following in inequality, known as the **Cauchy–Schwarz inequality**:

$$[E(XY)]^2 \le E(X^2)E(Y^2)$$

Use this inequality to show that r^2, the squared correlation coefficient, is such that $0 \le r^2 \le 1$.

1.9 Show that

a. $\bar{e} = 0$, that is, the average value of the residuals is zero.

b. $\bar{Y} = b_1 + b_2 \bar{X}_2 + b_3 \bar{X}_3 + \cdots + b_k \bar{X}_k$, that is, the regression hyperplane passes through the sample mean values of Y and the Xs.

c. The mean value of actual Y and estimated Y values are the same.

1.10 Consider the following regression model:

$$Y_i^k = B_1 + B_2 X_i + u_i \quad Y > 0, k = \text{a constant}$$

Consider the following values for k

$$k = 1, 2, 0.5, -0.5, -1$$

a. For each of these k values, find the corresponding regression model. Which of these models are LRMs?

b. Suppose $k=0$. Can you estimate the regression model in this case?[10]

1.11 You are given 10 values for variables Y (the dependent variable) and X (the explanatory variable):

Y 70 65 90 95 110 115 120 140 155 150

X 80 100 120 140 160 180 200 220 240 260

Based on these values, estimate the following regression:

$$Y_i = B_1 + B_2 X_i + u_i \quad i = 1, 2, \ldots, 10$$

a. Find $(X'X), (X'Y)$.

b. Estimate $b = (X'X)^{-1} X'Y$.

c. Estimate R^2 for this example.

---

[10]On this, see Box, G. E. P., & Cox, D. R. (1964). An analysis of transformations. *Journal of the Royal Statistical Society, B26,* 211–243.

Appendix 1A: Derivation of the Normal Equations

We start with the following equation:

$$\Sigma u_i^2 = \Sigma (Y_i - B_1 - B_2 X_{2i} - B_3 X_{3i} - \cdots - B_k X_{ki})^2 \tag{1A.1}$$

Differentiating this equation partially with respect to each of the B coefficients, and setting them equal to 0, we obtain

$$\frac{\partial \Sigma u_i^2}{\partial B_1} = 2\Sigma (Y_i - B_1 - B_2 X_{2i} - B_3 X_{3i} - \cdots - B_k X_{ki})(-1) = 0$$

$$\frac{\partial \Sigma u_i^2}{\partial B_2} = 2\Sigma (Y_i - B_1 - B_2 X_{2i} - B_3 X_{3i} - \cdots - B_k X_{ki})(-X_{2i}) = 0$$

$$\frac{\partial \Sigma u_i^2}{\partial B_3} = 2\Sigma (Y_i - B_1 - B_2 X_{2i} - B_3 X_{3i} - \cdots - B_k X_{ki})(-X_{3i}) = 0 \tag{1A.2}$$

$$\vdots$$

$$\frac{\partial \Sigma u_i^2}{\partial B_k} = 2\Sigma (Y_i - B_1 - B_2 X_{2i} - B_3 X_{3i} - \cdots - B_k X_{ki})(-X_{ki}) = 0$$

We have k equations in k unknown B coefficients and *in general* we can obtain the values of the k unknown regression parameters.[11] Replacing the unknown, and unobservable, B by the observable (or computed) b, and simplifying, we obtain

$$nb_1 + b_2 \Sigma X_{2i} + b_3 \Sigma X_{3i} + \cdots + b_k \Sigma X_{ki} = \Sigma Y_i$$

$$b_1 \Sigma X_{2i} + b_2 \Sigma X_{2i}^2 + b_3 \Sigma X_{2i} X_{3i} + \cdots + b_k \Sigma X_{2i} X_{ki} = \Sigma X_{2i} Y_i$$

$$b_1 \Sigma X_{3i} + b_2 \Sigma X_{3i} X_{2i} + b_3 \Sigma X_{3i}^2 + \cdots + b_k \Sigma X_{3i} X_{ki} = \Sigma X_{3i} Y_i \tag{1A.3}$$

$$b_1 \Sigma X_{ki} + b_2 \Sigma X_{ki} X_{2i} + b_3 \Sigma X_{ki} X_{3i} + \cdots + b_k \Sigma X_{ki}^2 = \Sigma X_{ki} Y_i$$

Notice that each estimated B coefficient is expressed in terms of the observable squares and cross-products of the X and Y variables.

---

[11]We say in general because it is quite possible that these equations are not independent of each other. In Chapter 2, we will discuss this problem more carefully.

In matrix notation, Equation (3) can be expressed as

$$\begin{pmatrix} n & \Sigma X_{2i} & \Sigma X_{3i} & \cdots\cdots & \Sigma X_{ki} \\ \Sigma X_{2i} & \Sigma X_{2i}^2 & \Sigma X_{2i}X_{3i} & \cdots\cdots & \Sigma X_{2i}X_{ki} \\ \Sigma X_{3i} & \Sigma X_{3i}X_{2i} & \Sigma X_{3i}^2 & \cdots\cdots & \Sigma X_{3i}X_{ki} \\ & & & & \\ & & & & \\ \Sigma X_{ki} & \Sigma X_{ki}X_{2i} & \Sigma X_{ki}X_{3i} & \cdots\cdots & \Sigma X_{ki}^2 \end{pmatrix} \begin{bmatrix} b_1 \\ b_2 \\ b_3 \\ \\ \\ b_k \end{bmatrix} = \begin{bmatrix} 1\ 1\ 1\ 1 \cdots\cdots & 1 \\ X_{21}X_{22}X_{23}\cdots\cdots & X_{2n} \\ X_{31}\ X_{32}\ X_{33}\cdots\cdots & X_{3n} \\ \\ X_{k1}\ X_{k2}\ X_{k3}\cdots\cdots X_{kn} \end{bmatrix} \begin{bmatrix} Y_1 \\ Y_2 \\ Y_3 \\ \\ \\ Y_n \end{bmatrix}$$

$$(X'X) \qquad\qquad b \qquad\qquad X' \qquad\qquad y$$

$$(k \times k)(k \times 1)(k \times n)(n \times 1) \tag{1A.4}$$

Verify that the matrices and vectors follow the rules of matrix algebra. More compactly, Equation (4) can be expressed as

$$(X'X)b = X'y \tag{1A.5}$$

Since X is of rank k, and the inverse of $(X'X)$ exists, denoted by $(X'X)^{-1}$, we can premultiply Equation (5) on both sides by this inverse to obtain

$$(X'X)^{-1}(X'X)b = (X'X)^{-1}(X'y) \tag{1A.6}$$

Since $(X'X)^{-1}(X'X) = I$, an identity matrix of order $(n \times n)$, we obtain

$$b = (X'X)^{-1}(X'y) \tag{1A.7}$$
$$(k \times 1)(k \times k)(k \times n)(n \times 1)$$

This is a fundamental result of OLS regression.

It may be noted that b given in Equation (1A.7) is an example of a **linear estimator**.

It is of the form Ly, where L is a matrix of real numbers; in the present case,

$$L = (X'X)^{-1}X' \tag{1A.8}$$

CHAPTER 2. THE CLASSICAL
LINEAR REGRESSION MODEL (CLRM)

In Chapter 1, we showed how we estimate an LRM by the method of least squares. As noted in Chapter 1, estimation and hypothesis testing are the twin branches of statistical inference. Based on the OLS, we obtained the sample regression, such as the one shown in Equation (1.40). This is of course a sample regression function (SRF) because it is based on a specific sample drawn randomly from the purported population. What can we say about the true population regression function (PRF) from the SRF? In practice, we do not observe the PRF and have to "guess" it from the SRF. To obtain the best possible guess, we need a framework, which is provided by the **classical linear regression model (CLRM)**. The CLRM is based on several assumptions, which are discussed below.

2.1 Assumptions of the CLRM

We now discuss these assumptions. In Chapters 5 and 6, we will examine these assumptions more critically. However, keep in mind that in any scientific inquiry we start with a set of simplified assumptions and gradually proceed to more complex situations.

Assumption 1: The regression model is *linear in the parameters* as in Equation (1.1); it may or may not be linear in the variables, the Ys and Xs.

Assumption 2: The regressors are assumed fixed, or nonstochastic, in the sense that their values are fixed in repeated sampling. However, if the regressors are stochastic, we assume that each regressor is independent of the error term or at least uncorrelated with it. We will discuss this assumption in more detail in Chapter 6.

Assumption 3: Given the values of the X variables, the expected, or mean, value of the error term u_i is 0.

$$E(u_i \mid X) = 0 \qquad (2.1)$$

In matrix notation, we have

$$E(u \mid X) = 0 \qquad (2.1a)$$

where 0 is the **null vector**.

More explicitly,

$$E \begin{bmatrix} u_1 \\ u_2 \\ u_3 \\ \vdots \\ u_n \end{bmatrix} = \begin{bmatrix} E(u_1) \\ E(u_2) \\ E(u_3) \\ \vdots \\ E(u_n) \end{bmatrix} = \begin{bmatrix} 0 \\ 0 \\ 0 \\ \vdots \\ 0 \end{bmatrix}$$

Because of this *critical* assumption, and given the values of the regressors, we can write Equation (1.5) as

$$E(y \mid X) = BX + E(u \mid X)$$
$$= BX \qquad (2.2)$$

This is the PRF. In regression analysis, our primary objective is to estimate this function. The PRF thus gives the mean value of the regressand corresponding to the given values of the regressors, noting that conditional on these values the mean value of the error term is 0.

Assumption 4: The variance of each u_i, given the values of X, is constant or **homoscedastic** (i.e., of equal variance). That is,

$$\text{var}(u_i \mid X) = \sigma^2 \qquad (2.3)$$

In matrix notation,

$$\text{var}(u \mid X) = E(uu') = \sigma^2 I \qquad (2.3a)$$

where I is an $n \times n$ identity matrix (see also Assumption 5).

If $\text{var}(u_i \mid X) = \sigma_i^2$, the error variance is said to be heteroscedastic, or of unequal variance. We will discuss this case in Chapter 5.

Figure 2.1 is a picture of both homoscedasticity and heteroscedasticity.

Assumption 5: There is no correlation between error terms belonging to two different observations. That is,

$$\text{cov}(u_i, u_j \mid X) = 0, \quad i \neq j \qquad (2.4)$$

where cov stands for covariance, and i and j are two different error terms. Of course, if $i = j$, we get the variance of u_i given in Equation (2.3).

Figure 2.2 shows a likely pattern of autocorrelation.

Figure 2.1 Homoscedasticity and Heteroscedasticity

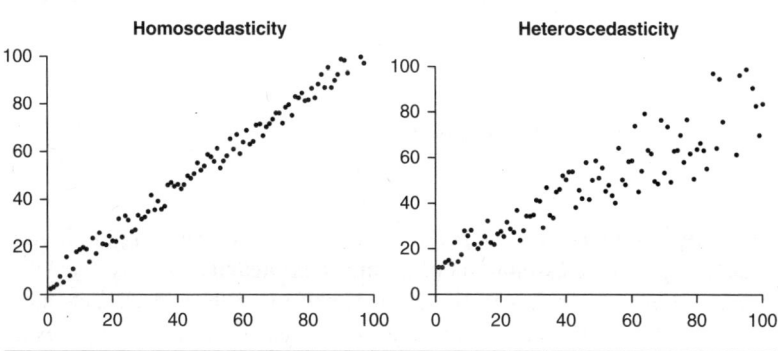

Figure 2.2 Autocorrelation

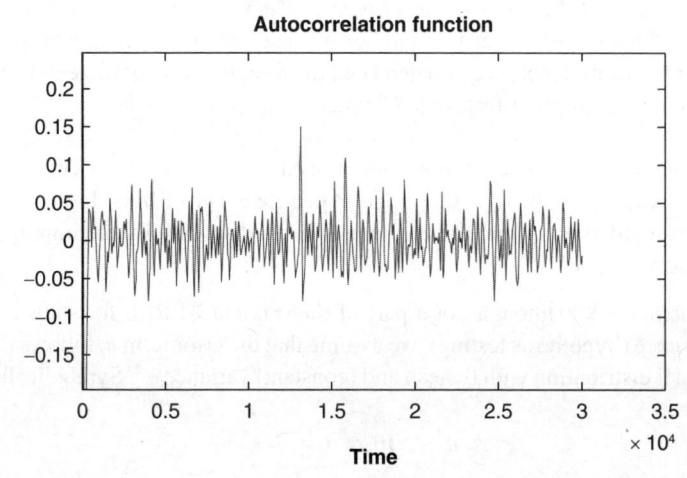

Assumptions 4 and 5 can be expressed as

$$E(uu') = \sigma^2 I \quad \text{if } i = j$$
$$= 0 \quad \text{if } i \neq j$$

where **0** is the **null matrix** and I is the identity matrix. We discuss this assumption further in Chapter 5. More compactly, we can express Assumptions 4 and 5 as

$$E(\boldsymbol{uu'}) = \sigma^2 \boldsymbol{I} = \begin{pmatrix} \sigma^2 & \cdots & 0 \\ \vdots & \ddots & \vdots \\ 0 & \cdots & \sigma^2 \end{pmatrix}$$

Assumption 6: There is *no perfect linear relationship* among the X variables. This is the assumption of no **multicollinearity**. Strictly speaking, *multicollinearity* refers to the existence of more than one exact linear relationship, and *collinearity* refers to the existence of a single exact linear relationship. But this distinction is rarely maintained in practice, and multicollinearity refers to both cases. Imagine what would happen in the wage regression given in Equation (1.5), if we were to include work experience both in years and in months!

In matrix notation, this assumption means that the X matrix is of *full column rank*. In other words, *the columns of the* X *matrix are linearly independent*. This requires that the number of observations, n, is greater than the number of parameters estimated (i.e., the k regression coefficients). We discuss this assumption further in Chapter 7.

Assumption 7: The regression model used in the analysis is **correctly specified**, that is, there is no (model) **specific error** or **bias**. In practice, this is a tall assumption, but in Chapter 7, we discuss fully the import of this assumption.

Assumption 8: Although not a part of the original CLRM, for statistical inference (hypothesis testing), we assume that the error term u_i follows the normal distribution with 0 mean and (constant) variance σ^2. Symbolically,

$$u_i \sim N(0, \sigma^2) \tag{2.5}$$

Or in matrix notation,

$$\boldsymbol{u} \sim N(\boldsymbol{0}, \sigma^2 \boldsymbol{I}) \tag{2.5a}$$

The assumption of the normality of the error term is crucial if the sample size is rather small; it is not essential if we have a very large sample. However, we will revisit this assumption in Chapter 7. With this assumption, CLRM is known as the **classical normal linear regression model (CNLRM)**.

Since we are assuming that the X matrix is nonstochastic but u is stochastic, the regressand Y is also stochastic. In addition, since u is normally distributed with 0 mean and constant variance, Y inherits the properties of u. More specifically,

$$y \sim N(BX, \sigma^2 I) \tag{2.6}$$

That is, the regressand is distributed normally with mean BX and the (constant) variance σ^2.

Under Assumption 8, we can use the method of **maximum likelihood (ML)** as an alternative to OLS. We will discuss ML more thoroughly in Chapter 3 because of its general applicability in many areas of statistics.

With one or more of the preceding assumptions, in this chapter, we discuss the following topics:

1. The sampling distribution of the OLS estimators, b

2. An estimator of the unknown variance, σ^2

3. The relationship between the residual e and the error u

4. Small-sample properties of the OLS estimators

5. Large-sample properties of the OLS estimators

2.2 The Sampling or Probability Distributions of the OLS Estimators

Remember that the population parameters in B, although unknown, are constants. However, this is not true of the estimated b coefficients, for their values depend on the sample data at hand. In other words, the b coefficients are random. As such, we would like to find their sampling or probability distributions to establish properties of the (OLS) estimators.

Recall that

$$b = (X'X)^{-1} X'y, \quad \text{using Equation } (1.16)$$

Therefore,

$$b = (X'X)^{-1} X' [(XB + u)], \quad \text{using Equation } (1.5)$$
$$= (X'X)^{-1} X'XB + (X'X)^{-1} X'u$$
$$= B + (X'X)^{-1} X'u \tag{2.7}$$

By the definition of covariance, we obtain

$$\text{cov}(b) = E(b - B)(b - B)' = E\left[(X'X)^{-1}X'u\right]\left[(X'X)^{-1}X'u\right]',$$

using Equation (2.7)

$$= E\left[(X'X)^{-1}X'uu'\,X(X'X)^{-1}\right]$$

$$= (X'X)^{-1}X'E(uu')X(X'X)^{-1}$$

$$= (X'X)^{-1}X'\sigma^2 IX(X'X)^{-1}$$

$$= \sigma^2(X'X)^{-1} \tag{2.8}$$

In deriving this expression, we have used properties of the transpose of an inverse matrix, and the assumption that X is fixed and that the variance of u_i is constant and the us are uncorrelated. *Notice that we can move the expectations operator through* $\mathbf{X}$ *because it is assumed fixed.* The variances of the individual elements of b are on the main diagonal (running from upper left to lower right), and the off-diagonal elements give the covariances between pairs of coefficients in b.

Since

$$b = (X'X)^{-1}X'y \quad (1.16)$$

and the X matrix is fixed, b is a linear function of y. Using Assumption 8, we know that y is normally distributed. It is a property of the normal distribution that any linear function of a normally distributed variable is also normally distributed. Therefore, b is ipso facto normally distributed as follows:

$$b \sim N(B, \sigma^2(X'X)^{-1}) \tag{2.9}$$

That is, b is normally distributed with B as its mean (see Equation 2.20) and the variance established in Equation (2.8). In other words, under the normality assumption, the **sampling distribution** of the OLS estimator is normal, as shown in Equation (2.9). This finding will aid us in testing hypotheses about any element of B or any linear combination thereof. It may be noted that a sampling distribution is a probability distribution of an estimator or of any test statistic. In other words, it describes the variation in the values of a statistic over all possible samples, here the variation in b over all possible samples.[1]

---

[1]Suppose we draw several independent samples and for each sample we compute a (test) statistic, such as the mean, and draw a frequency distribution of all these statistics. Roughly speaking, this frequency distribution is the sampling distribution of that statistic. In our case, under the assumed conditions, the probability or sampling distribution of any component of b is normal as shown in Equation (2.9).

For any single element of b, b_k, we can express Equation (2.9) as

$$b_k \sim N(B_k, \sigma^2 x^{kk}) \tag{2.9a}$$

where x^{kk} is the kth diagonal element of $(X'X)^{-1}$. The square root of $\sigma^2 x^{kk}$ will give the standard error of b_k (see Figure 2.3).

However, before we can engage in hypothesis testing, we need to estimate the unknown σ^2. Remember that σ^2 refers to the variance of the error term u. Since we do not observe u directly, we have to rely on the estimated residuals, e, to learn about the true variance. Toward that end, we need to establish the relationship between u and e. Recall that

$$e = y - \hat{y} \tag{2.10}$$

Substituting for $\hat{y}$ from (1.23), we obtain

$$e = y - Xb$$
$$= y - X(X'X)^{-1}X'y, \text{ substituting for } b \text{ from Eq. (1.16)}$$
$$= My$$
$$= M(XB + u)$$

where

$$M = [I - X(X'X)^{-1}X']$$
$$= Mu, \text{ because } MXB = 0.^2 \tag{2.11}$$

Figure 2.3 The Distribution of b_k, a Component of the Vector b

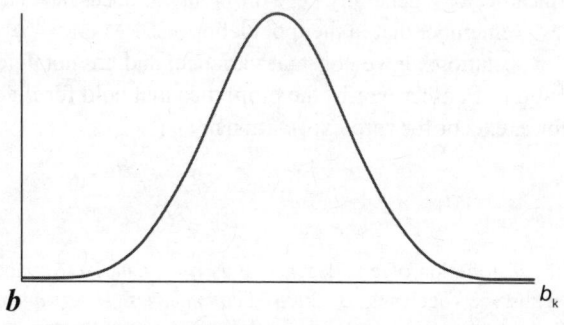

$^2 MXB = [XB - X(XX')^{-1}X'XB] = XB - IXB = 0$, where I is the identity matrix.

As noted in Chapter 1, M is a very important matrix in the analysis of LRMs. It is an **idempotent matrix**, a square matrix with the property that $M = M^2$. For further properties of the idempotent matrices, see Appendix A on linear algebra.

Since M is constant because it is a function of (fixed) X, we can write

$$E(e) = E(Mu)$$

$$= ME(u)$$

$$= 0 \qquad (2.12)$$

because $E(u) = 0$, by Assumption 1. We have thus shown that the expectation of each element of e is 0.

Now,

$$\text{cov}(e) = E(ee') = E(Muu'M')$$

$$= ME(uu')M'$$

$$= \sigma^2 IM = \sigma^2 M \qquad (2.13)$$

recalling the properties of M. This equation gives the covariance matrix of e.

Since e is a linear function of u and since u is normally distributed by Assumption 8, we have

$$e \sim N(0, \ \sigma^2 M) \qquad (2.14)$$

Therefore, like the mean of u, the mean of e is 0, but unlike u, the residuals are heteroscedastic as well as autocorrelated.[3]

What Equations (2.13) and (2.14) show is that the residuals $e_1, e_2, \ldots, e_n$ have zero mean values, generally have different variances, and have nonzero covariances. Remember that in the (population) CLRM errors, $u_1, u_2, \ldots, u_n$ have zero expectations, have constant variance, and are not autocorrelated (by assumption). In other words, the properties that hold for u generally do not hold for e, except for zero expectations.

---

[3]Actually, the distribution of e is degenerate as its variance–covariance matrix is singular. On this, see Vogelvang, B. (2005). *Econometrics: Theory and applications with Eviews* (chapter 4). Harlow, England: Pearson-Addison Wesley.

Although we have assumed that the variance of u (not of e) is constant, equal to σ^2, we are yet to estimate it from the sample data. Toward that end, we proceed as follows.

Even though we do not observe u, we observe e (after the regression is estimated). Naturally, we will have to estimate the unknown variance from the estimated e. From Equation (2.11), we know that

$$e = Mu \qquad (2.11)$$

Therefore,

$$
\begin{aligned}
E(e'e) &= E(u'M'Mu) \\
&= E(u'Mu) \qquad (2.15)
\end{aligned}
$$

because of the properties of M. Now,

$E(e'e) = E(u'Mu)$

$\qquad = E\,[\mathrm{tr}(u'Mu)]$, since $u'Mu$ is a scalar

$\qquad = E\,[\mathrm{tr}\,(Muu')]$ changing the order of multiplication inside the trace

$\qquad = \mathrm{tr}[M\,(Euu')]$, since the trace and expectations operators are both linear

$\qquad = \mathrm{tr}[M(\sigma^2 I)]$

$\qquad = \sigma^2 \mathrm{tr}(M)$

$\qquad = \sigma^2 \left[\mathrm{tr}(I) - \mathrm{tr}(X(X'X)^{-1}X')\right]$, using the definition of M

$\qquad = \sigma^2 \left[\mathrm{tr}(I_n) - \mathrm{tr}((X'X)^{-1}X'X)\right]$

$\qquad = \sigma^2 \left[\mathrm{tr}(I_n) - \mathrm{tr}(I_k)\right]$, since $(X'X)^{-1}X'X = I_k$

$\qquad = \sigma^2 (n-k)$, since $\mathrm{tr}(I) = n$ and $\mathrm{tr}(I_k) = k \qquad (2.16)$

The notation $\mathrm{tr}(M)$ means the trace of the matrix M, which is simply the sum of the entries of the main diagonal of M. In deriving the steps in Equation (2.16), we have made use of several properties of the *trace* of a matrix, such as the fact that trace is a linear operator and if AB and BA are both square matrices, then $\mathrm{tr}(AB) = \mathrm{tr}(BA)$.

As a result, we can now write

$$E\left(\frac{e'e}{n-k}\right)=\sigma^2 \qquad (2.17)$$

If we now define

$$S^2 = \frac{e'e}{n-k} = \frac{\Sigma e_i^2}{n-k} \qquad (2.18)$$

then

$$E(S^2)=\sigma^2 \qquad (2.19)$$

In words, S^2 is an unbiased estimator of the true error variance σ^2. S, the square root of S^2, is called the **standard error (se) of the estimate** or the **standard error of the regression**. In practice, therefore, we use S^2 in place of σ^2.

2.3 Properties of OLS Estimators: The Gauss–Markov Theorem[4]

The OLS estimators possess some ideal or optimum properties, which are contained in the well-known **Gauss–Markov theorem:**[5] Given the assumptions of the classical regression model, in the class of unbiased linear estimators, the least-squares estimators have minimum variance; that is, they are **best linear unbiased estimators, BLUE** for short. In other words, no other linear, unbiased estimator of B can have a smaller variance than the OLS estimator given in Equation (2.8).

To establish this theorem, first note that b, the OLS estimator of B, is a linear function of the regressand y, as we have established in Chapter 1 (see Equation 1.16).[6] To prove that b is unbiased, we proceed as follows:

$$b=(X'X)^{-1}X'y \qquad (1.16)$$
$$=(X'X)^{-1}X'[XB+u], \text{ substituting for } y$$
$$=(X'X)^{-1}X'XB+(X'X)^{-1}X'u$$
$$=B+(X'X)^{-1}X'u$$

---

[4]In Appendix C, we discuss both small-sample and large-sample properties of OLS and ML estimators.

[5]Although known as the *Gauss–Markov theorem*, the least-squares approach of Gauss antedates (1821) the minimum-variance approach of Markov (1900).

[6]See the discussion in Darnell, A. C. (1994). *A dictionary of econometrics* (p. 155). Cheltenham, England: Edward Elgar.

Now

$$E(b) = B + (X'X)^{-1} X' E(u)$$
$$= B \qquad (2.20)$$

In words, the expected value of **b** is equal to **B**, thus proving that **b** is unbiased. (Recall the definition of unbiased estimator.) Note that $E(u|X)=0$ by assumption.

To prove that in the class of unbiased linear estimators the least-squares estimators have the least variance (i.e., they are efficient), we proceed as follows:

Let b^* be another linear estimator of **B** such that

$$b^* = [A + (X'X)^{-1} X'] y \qquad (2.21)$$

where A is some nonstochastic $k \times n$ matrix, similar to X. Simplifying, we obtain

$$b^* = Ay + (X'X)^{-1} X' y$$
$$= Ay + b \qquad (2.22)$$

where **b** is the least-squares estimator given in Equation (1.16).

Now

$$E\left(b^*\right) = \left[A + (X'X)^{-1} X'\right] E\left(y\right)$$
$$= \left[A + (X'X)^{-1} X'\right] (XB)$$
$$= (AX + I) B \qquad (2.23)$$

Now $E(b^*) = B$ if and only if $AX = 0$. In other words, for the linear estimator b^* to be unbiased, AX must be **0**.

Thus,

$$b^* = \left[A + (X'X)^{-1} X'\right] [XB + u], \quad \text{substituting for } (y)$$
$$= B + \left[A + (X'X)^{-1} X'\right] u, \quad \text{because } AX = 0$$

Given that **u** has zero mean and constant variance ($= \sigma^2 I$), we can now find the variance of b^* as follows:

$$\text{cov}(b^*) = E\left[A + (X'X)^{-1} X'\right] uu' \left[A + (X'X)^{-1} X'\right]'$$
$$= \left[A + (X'X)^{-1} X'\right] E(uu') \left[A + (X'X)^{-1} X'\right]'$$
$$= \sigma^2 \left[AA' + (X'X)^{-1}\right]$$
$$= \sigma^2 (X'X)^{-1} + AA' \sigma^2$$
$$= \text{var}\left(b\right) + AA' \sigma^2 \qquad (2.24)$$

Since AA' is a positive semidefinite matrix, Equation (2.24) shows that the covariance matrix of b^* is equal to the covariance matrix of b plus a positive semidefinite matrix. That is, $\text{cov}(b^*) > \text{cov}(b)$, unless $A = 0$. This shows that in the class of unbiased linear estimators, the least-square estimator b has the least variance, that is, it is efficient compared with any other linear unbiased estimator of B.

It is important to note that in establishing the Gauss–Markov theorem we do not have to assume that the error term u follows a particular probability distribution, such as the normal. To establish the theorem, we only need Assumptions 1 to 5.

It is also important to note that if one or more assumptions underlying the Gauss–Markov theorem are not satisfied, the OLS estimators will not be BLUE. Also, bear in mind that the Gauss–Markov theorem holds only for linear estimators, that is, linear functions of the observation vector y. There are situations where nonlinear (in-the-parameter) estimators are more efficient than the linear estimators. In this book, we do not deal with nonlinear estimators, for that requires a separate book.[7]

To sum up, we have shown that under the Gauss–Markov assumptions, b, the least-square estimator of B, is BLUE, that is, in the class of unbiased linear estimators, b has the least variance. We also showed how to estimate B and the variance of the estimated B.

2.4 Estimating Linear Functions of the OLS Parameters

We have shown how to estimate B, that is, each of its elements. Suppose we want to estimate some linear function of the elements of B, that is, that of $B_1, B_2, B_3, \ldots, B_k$. More specifically, suppose we want to estimate $t'B$, where t' is a $1 \times k$ vector of real numbers and B is a $k \times 1$ vector of the parameters in B. It can be shown that the BLUE of $t'B$ is $t'b$, where b is the least-square estimator of B (see also Appendix C).

What this means is that whether we estimate all the elements of B, or one of its elements, or estimate a linear combination ($t'B$), we can use the OLS regression.

Let $\lambda = t'B$. By choosing t appropriately, we can make λ equal to any element of B, or to the sum of the elements of B that might be of interest to researchers. Suppose in Equation (1.2), we want the coefficient of B_1 equal to 4, and the coefficient of B_5 equal to -1, and the rest of the coefficients to be all zeros. In other words, we want $\lambda = 4B_1 - B_5$. Here, $t' = (4, 0, 0, 0, -1, 0, 0, 0, \ldots)'$.

Using the definition of variance, we can now find the variance of the estimated λ ($= \hat{\lambda}$), which is

---

[7]For examples of nonlinear estimators and their applications, see Gujarati, D. (2015). *Econometrics by example* (2nd ed.). London, England: Palgrave Macmillan.

$$\text{var}\,(\hat{\lambda}) = t'(\text{var}(\boldsymbol{b}))t = \sigma^2 t'(\boldsymbol{X'X})^{-1}t \qquad (2.25)$$

But keep in mind that in general the variance of $\hat{\lambda}$ depends on every element of the covariance matrix of $\boldsymbol{b}$, the estimator of $\boldsymbol{B}$. However, if some elements of the vector t are equal to zero, var($\hat{\lambda}$) does not depend on the corresponding rows and columns of the covariance matrix $\sigma^2(\boldsymbol{X'X})^{-1}$.

As an example, consider $\lambda = 4B_1 - B_5$. In this case,

$$\begin{aligned}
\text{var}\,(\hat{\lambda}) &= t_1^2 \,\text{var}(b_1) + t_5^2 \,\text{var}(b_5) + 2t_1 t_5 \,\text{cov}(b_1, b_5) \\
&= 16\,\text{var}(b_1) + \text{var}(b_5) - 8\,\text{cov}(b_1, b_5)
\end{aligned} \qquad (2.26)$$

Notice in this example only the variances of b_1 and b_5 and their covariances are involved, as the values of the other parameters in the k-variable regression (1.2) are assumed to be zero. But if there are more nonzero coefficients, the variances and their pairwise covariances will also be involved in computing the variance of the linear combination $t'\boldsymbol{B}$.

2.5 Large-Sample Properties of OLS Estimators

2.5.1 Consistency of OLS Estimators

We have shown that the OLS estimators of the CLRM are unbiased, which is a small, or finite sample, property. We can also show that the OLS estimators are consistent, that is, they converge to their true values as the sample size increases indefinitely. Convergence is a large-sample property.

Proof: A sufficient condition for an unbiased estimator to be consistent is for its variance to converge to zero as the sample size n increases indefinitely. For the OLS estimator $\boldsymbol{b}$, we have already shown that its variance is

$$\text{cov}(\boldsymbol{b}) = \sigma^2 (\boldsymbol{X'X})^{-1} \qquad (2.8)$$

which we can write as

$$\text{cov}(\boldsymbol{b}) = \frac{\sigma^2}{n}(n^{-1}\boldsymbol{X'X})^{-1} \qquad (2.27)$$

To see the behavior of this expression as $n \to \infty$, we have

$$\begin{aligned}
p\lim_{n \to \infty} \text{cov}(\boldsymbol{b}) &= p\lim_{n \to \infty}\left[\frac{\sigma^2}{n}(n^{-1}\boldsymbol{X'X})^{-1}\right] \\
&= p\lim_{n \to \infty}\frac{\sigma^2}{n}\lim_{n \to \infty}(n^{-1}\boldsymbol{X'X})^{-1}
\end{aligned} \qquad (2.28)$$

where $p\lim$ is probability limit (see Appendix C for details).

We have assumed that the elements of the matrix X are bounded, which means the second term in the preceding equation is bounded for all n. Therefore, the second term above can be replaced by a matrix of finite constants. Now, the limit of the first term in Equation (2.28) tends to zero as n increases indefinitely. As a result,

$$\plim_{n \to \infty} \text{cov}(b) = 0 \tag{2.29}$$

which establishes that b is a consistent estimator of B. In establishing this result, we have used some of the properties of the $\plim$ operator.

2.5.2 Consistency of the OLS Estimator of the Error Variance

We have proved that S^2 is an unbiased estimator of σ^2. Assuming values of u_i are independent and identically distributed (iid), we can prove that S^2 is also a consistent estimator of σ^2. The proof is as follows:

$$
\begin{aligned}
S^2 &= \frac{(y - Xb)'(y - Xb)}{n - k} \\
&= \frac{u'(I - X(X'X)^{-1}X')u}{n - k} \\
&= \left(\frac{n}{n-k} \right) \left(\frac{u'u}{n} - \frac{u'X}{n} \cdot \left(\frac{X'X}{n} \right)^{-1} \cdot \frac{X'u}{n} \right)
\end{aligned}
\tag{2.30}
$$

Note: $e = My = Mu$, where $M = [I - X(X'X)X']$. Also note how the entries are manipulated by multiplying or dividing them by the sample size or the adjusted sample size without affecting the basic relationships.

Taking the $\plim$ of both sides of Equation (2.30), we obtain

$$
\begin{aligned}
\plim S^2 &= \plim \left(\frac{n}{n-k} \right) \left(\plim \frac{u'u}{n} - \plim \frac{u'X}{n} \cdot \plim \left(\frac{X'X}{n} \right)^{-1} \cdot \plim \frac{X'u}{n} \right) \\
&= 1(\sigma^2 - 0 \cdot Q^{-1} \cdot 0), \quad \text{where} \plim \left(\frac{X'X}{n} \right)^{-1} = Q^{-1} \\
&= \sigma^2
\end{aligned}
\tag{2.31}
$$

which establishes the result. Note that for large n, $(n - k) \approx n$.

In deriving the preceding result, we have used **Khinchine's theorem** (see Appendix B) as well as the properties of the $p$lim.

2.5.3 Independence of the OLS
Estimators and the Residual Term, e

What this says is that each element of b is uncorrelated with each element of the least-squares residual vector e. The proof is as follows:

Recall that

$$b = B + (X'X)^{-1}X'u \quad (2.7)$$

$$e = Mu \quad (2.11)$$

are both linear functions of the error term u.

Now the covariance of b and e is

$$\text{cov}(b,e) = (X'X)^{-1}X'\text{var}(u)M'$$
$$= \sigma^2(X'X)^{-1}X'M'$$
$$= 0, \text{ since } MX = 0 \leftrightarrow X'M' \quad (2.32)$$

This shows that b and e are uncorrelated.

It may be noted that if we assume that u is normally distributed, b and e are not only uncorrelated but also independent. In Chapter 3, we will consider the **normal linear regression model**, which explicitly assumes that the error term u is normally distributed and will see the consequence of the normality assumption.

2.5.4 Large-Sample Distribution of b:
Asymptotic Normality of the OLS Estimators

It can be shown that[8]

$$b\,\text{asy} \sim N(B, \sigma^2(X'X)^{-1}) \quad (2.33)$$

where asy means asymptotically (i.e., $n \to \infty$).

In other words, as the sample size n increases indefinitely, b is approximately normally distributed with mean equal to B and variance equal to

---

[8]The proof is rather complicated and can be found in Theil, H. (1971). *Principles of econometrics* (pp. 380–381). New York, NY: Wiley; see also Mittlehammer, R. C. (1996). *Mathematical statistics for economics and business* (pp. 443–447). New York, NY: Springer.

$\sigma^2(X'X)^{-1}$. Each element of b is individually normally distributed with variance equal to the appropriate element of the variance matrix $\sigma^2(X'X)^{-1}$. *This result holds whether* u *is normally distributed or not.*

It may be noted that if the errors u_i are not iid, even then b is asymptotically normally distributed as in (2.33) under certain conditions.[9]

2.5.5 Asymptotic Normality of S^2

If in addition to the classical assumptions, it is assumed that values of u_i are iid and have bounded fourth-order moments about the origin, S^2 is asymptotically normally distributed. These results also hold true even if the u_i values are not iid.[10]

2.6 Summary

The CLRM, $y=XB+u$, is the foundation of regression analysis. It is based on several assumptions. The basic assumptions are that (1) the data matrix X is nonstochastic, (2) it is of full column rank, (3) the expected value of the error term is zero, and (4) the covariance matrix of the error term $E(uu')=\sigma^2 I$. This means the error variance is constant and equal to σ^2 and that the error terms are mutually uncorrelated.

We used the method of OLS to estimate the parameters of an LRM. One reason for using OLS is that it does not require us to make assumptions about the probability distribution of the error term, and it is comparatively easy to estimate. Parameters of the CLRM estimated by OLS are called OLS estimators. OLS estimators have several desirable statistical properties such as (1) they are unbiased and (2) among all linear unbiased estimators of B, they have minimum variances. This is called the Gauss–Markov theorem. These are small-sample properties.

OLS estimators have these asymptotic, or large-sample, properties: (1) The OLS estimators of B as well as the estimator of the error variance are consistent estimators and (2) the OLS estimators asymptotically follow the normal distribution.

Exercises

2.1 Consider the bivariate regression: $Y_i = B_1 + B_2 X_i + u_i$. Under the classical linear regression assumptions, show that

---

[9]See Mittlehammer, R. C. (1996). *Mathematical statistics for economics and business* (p. 445). New York, NY: Springer.

[10]See Mittlehammer, R. C. (1996). *Mathematical statistics for economics and business* (pp. 448–449). New York, NY: Springer.

a. $\mathrm{cov}(b_1,b_2) = -\bar{X}\dfrac{\sigma^2}{\Sigma(X_i - \bar{X})^2}$

b. $\mathrm{cov}(\bar{Y},b_2) = 0$

2.2 Show that for the model in Exercise 2.1,

$$\mathrm{RSS} = \frac{\Sigma x_i^2 \Sigma y_i^2 - (\Sigma x_i y_i)^2}{\Sigma x_i^2}$$

where RSS is the residual sum of squares and

$$x_i = (X_i - \bar{X}); \quad y_i = (Y_i - \bar{Y}); \quad x_i y_i = (X_i - \bar{X})(Y_i - \bar{Y})$$

2.3 Verify the following properties of OLS estimators:

a. The OLS regression line (plane) passes through the sample means of the regressand and the regressors.

b. The mean values of the actual Y and the estimated $Y(=\hat{Y})$ are the same.

c. In the CLRM with intercept, the mean value of the residuals $(\bar{e})$ is zero.

d. As a result of the preceding property, the k-variable sample CLRM can be expressed as

$$y_i = b_2 x_{2i} + b_3 x_{3i} + \cdots + b_k x_{ki} + e_i$$

where $y_i = (Y_i - \bar{Y}); x_{ki} = (X_{ki} - \bar{X}_k)$

2.4 Consider the following bivariate regression model:

$$Y_i^* = B_1^* + B_2^* X_i^* + u_i$$

where

$$Y_i^* = \frac{Y_i - \bar{Y}}{s_Y}; \quad X_i^* = \frac{X_i - \bar{X}}{s_X}$$

where s_Y and s_X are the sample standard deviations of Y and X. Y_i^* and X_i^* are known as **standardized variables**, often known as **Z** scores. Since the units of measurement of the Z scores in the numerator and the denominator are the same, they are called "pure" or "unitless" numbers.

a. Show that a standardized variable has a zero mean and unit variance.

b. What are the formulas to estimate B_1^* and B_2^*?

c. What is the relationship between B_1^* and B_1 and between B_2^* and B_2?

2.5 The sample correlation coefficient between variables Y and X, r_{XY}, is defined as

$$r_{XY} = \frac{\Sigma x_i y_i}{\sqrt{\Sigma x_i^2 \Sigma y_i^2}}$$

where

$$x_i = (X_i - \bar{X}); \quad y_i = (Y_i - \bar{Y})$$

If we standardize variables as in Exercise 2.4, does it affect the correlation coefficient between X and Y? Show the necessary calculations.

2.6 Consider variables X_1, X_2, and X_3. Now consider the following correlation coefficients:

$$r_{12} = \text{correlation coefficient between } X_1 \text{ and } X_2$$
$$r_{13} = \text{correlation coefficient between } X_1 \text{ and } X_3$$
$$r_{23} = \text{correlation coefficient between } X_2 \text{ and } X_3$$

$$r_{12.3} = \frac{r_{12} - r_{13}r_{23}}{\sqrt{1 - r_{13}^2}\sqrt{1 - r_{23}^2}}$$

$r_{12.3}$ is called the **partial correlation coefficient** between X_1 and X_2 holding the influence of the variable X_3. The concept of partial correlation is akin to the concept of a partial regression coefficient.

a. What other partial correlation coefficients can you compute?

b. If we standardize the three variables as in Exercise 2.4, would the correlation coefficients among the standardized variables be different from the unstandardized variables?

c. Would partial correlation coefficients be affected by standardizing the variables? Explain.

2.7 Consider the following LRM:

$$Y_i = B_1 + B_2 X_{2i} + B_3 X_{3i} + B_4 X_{4i} + B_5 X_{5i} + u_i$$

How would you test the following hypotheses?

    a. $B_2 = B_3 = B_4 = B_5 = B$, that is, all partial regression coefficients are the same.

    b. $B_2 = B_3$ and $B_4 = B_5$

    c. $B_2 + B_3 = 2B_4$

2.8 Remember that the hat matrix, H, is expressed as

$$H = X(X'X)^{-1}X$$

Show that the residual vector e can also be expressed as

$$e = (I - H)y$$

2.9 Prove that the matrices H and $(I - H)$ are idempotent.

2.10* For the following matrix, compute its eigenvalues:

$$\begin{bmatrix} 1 & 0 & 0 \\ 0 & 1 & 0 \\ 0 & 0 & 1 \end{bmatrix}$$

(*Optional)

2.11 Consider the following regression model (see Chapter 7, Equation (7.30)):

$$Y_i = B_1 + B_2 X_i + B_3 X_i^2 + u_i$$

Models like this are called polynomial regression models, here a second-degree polynomial.

    a. Is this an LRM?

    b. Can OLS be used to estimate the parameters of this model?

    c. Since X_i^2 is the square of X_i, does this model suffer from perfect collinearity?

2.12 Consider the following model:

$$Y_i = B_1 + B_2 X_{2i} + B_3 X_{3i} + B_4 X_{4i} + u_i$$

You are told that $B_2 = 1$.

a. In this case, is it legitimate to estimate the following regression?

$$(Y_i - X_{2i}) = B_1 + B_3 X_{3i} + B_4 X_{4i} + u_i$$

This model is called a **restricted** linear regression, whereas the preceding model is called an **unrestricted** linear regression (see Chapter 4, Appendix 4A for further details).

b. How would you estimate the restricted regression, taking into account the restriction that $B_2 = 1$?

CHAPTER 3. THE CLASSICAL NORMAL LINEAR REGRESSION MODEL: THE METHOD OF MAXIMUM LIKELIHOOD (ML)

3.1 Introduction

We stated earlier that the assumption of the normal distribution for the error term u is not necessary if our objective is to obtain estimators of B only. As we have shown, OLS estimators of B did not require any distributional assumption about u. However, without making any assumption about the distribution of u, it is not possible to engage in statistical inference connecting b to B.

In the context of linear regression analysis, it is usually assumed that the error term u follows the normal distribution with zero mean and constant variance σ^2. The justification for the use of the normal distribution is as follows:

1. Since u represents all those variables that are not explicitly introduced in the model and whose individual influence is random and at best small, we can invoke the well-known **central limit theorem (CLT)** of statistics. The CLT states that if there are a large number of iid random variables, then, with a few exceptions, the distribution of their sum tends to a normal distribution as the number of such variables increases indefinitely (see Appendix B). The CLT therefore provides a justification for assuming that u is approximately normally distributed.

2. A version of the CLT states that, even if the number of variables is not very large, or if these variables are not strictly independent, their sum may still be normally distributed.[1]

3. With the normality assumption, the probability distributions of the OLS estimators can be derived easily. As we showed earlier, b, the OLS estimator of B, is a linear function of the regressand, y, which is a linear function of the error vector u; keep in mind that the data matrix X is fixed. If u is assumed to be normally distributed, which we do for statistical inference, then y, being a linear function of u, is also normally distributed. This is because a property of the normal

---

[1]For the various forms of the CLT, see Cramer, H. (1946). *Mathematical methods of statistics* (chap. 17). Princeton, NJ: Princeton University Press.

distribution is that *any linear function of a normally distributed variable is itself normally distributed*. Therefore, **b** is also normally distributed (see Equation 2.9).

4. There is a practical reason for using the normal distribution because it is a comparatively simple distribution involving only two parameters (mean and variance), it is very well-known, and statisticians have extensively studied its theoretical properties.

5. The normality assumption becomes very critical if the sample size is small, say, less than 50. It not only helps us in deriving the sampling distributions of the OLS estimators but also enables us to use the t, F, and chi-square (χ^2) probability distributions to test a variety of hypotheses. These probability distributions are related to the normal distribution. However, if the sample size is reasonably large, we will be able to relax the normality assumption.

With the normality assumption, we can develop an alternative to the method of OLS, which has several desirable statistical properties. This alternative is the **method of ML**.[2]

3.2 The Mechanics of ML

To ease the algebra, we will consider a bivariate, or two-variable, regression model:

$$Y_i = B_1 + B_2 X_i + u_i \tag{3.1}$$

where

$$u_i \sim \text{iid } N(0, \sigma^2) \tag{3.2}$$

That is, the error term is iid as a normal distribution with zero mean and constant variance σ^2.

Since B_1 and B_2 are constants and X is assumed fixed in repeated sampling, Y_i in Equation (3.1) is a linear function of u_i. Now if u_i is normally

[2]The ML method is of broader application and one can use a variety of distributional assumptions for the error term, not just the normal distribution. For example, the nonlinear logit model is based on the logistic probability distribution. On this, see Gujarati, D. (2015). *Econometrics by example* (2nd ed., chap. 8). New York, NY: Palgrave Macmillan.

distributed, Y_i, being a linear function of u_i, is also normally distributed, with mean and variance as follows:

$$Y_i \sim N(B_1 + B_2 X_i, \sigma^2) \qquad (3.3)^3$$

That is, the Y_i are normally and independently distributed with mean $= B_1 + B_2 X_i$ and variance $= \sigma^2$. Therefore, the *joint probability density function* (PDF) of $Y_1, Y_2, \ldots, Y_n$, with the stated mean and variance, can be expressed as

$$f(Y_1, Y_2, \ldots, Y_n) \mid (B_1 + B_2 X_i, \sigma^2) \qquad (3.4)$$

Since the Ys are assumed independent, this joint PDF can be written as a product of individual density function as

$$f(Y_1, Y_2, \ldots, Y_n) \mid (B_1 + B_2 X_i, \sigma^2) = f(Y_1 \mid Z) f(Y_2 \mid Z) \cdots f(Y_n \mid Z)$$

where $Z = (B_1 + B_2 X_i, \sigma^2)$.

Now the density function of a normally distributed Y_i can be written as (see Footnote 3):

$$f(Y_i) = \frac{1}{\sigma \sqrt{2\pi}} \exp\left[-\frac{1}{2\sigma^2} (Y_i - B_1 - B_2 X_i)^2 \right] \qquad (3.5)$$

Since each Y_i is independently distributed as in Equation (3.5), the joint density (i.e., the joint probability) of the Y observations can be written as the product of n such terms to obtain

$$LF = f(Y_1, Y_2, \ldots, Y_n) = \frac{1}{(\sqrt{\sigma^2})^n (\sqrt{2\pi})^n} \exp\left[-\frac{1}{2} \Sigma \frac{(Y_i - B_1 - B_2 X_i)^2}{\sigma^2} \right] \qquad (3.6)$$

If the Y values are known or given, but B_1, B_2, and σ^2 are unknown, the function (3.6) is called a **likelihood function**, denoted by $LF(B_1, B_2, \sigma^2)$.

The method of ML, as the name suggests, consists in estimating the unknown parameters in such a way that the probability of observing the sample of Y values is the maximum possible. Therefore, we have to find

---

[3]Recall from introductory statistics that the density function of a random normal variable X with mean μ and variance σ^2 is $f(X) = \frac{1}{\sigma \sqrt{2\pi}} \exp\left[-\frac{1}{2\sigma^2} (X - \mu)^2 \right]$, $-\infty < X < \infty, \sigma^2 > 0$.

the maximum of Equation (3.6). It is easy to find the maximum of Equation (3.6) if we take the (natural) logarithm of this function, which yields[4]

$$lF(B_1, B_2, \sigma^2) = -\frac{n}{2}\ln\sigma^2 - \frac{n}{2}\ln(2\pi) - \frac{1}{2}\Sigma\frac{(Y_i - B_1 - B_2 X_i)^2}{\sigma^2} \tag{3.7}$$

where l stands for the natural log of LF. Equation (3.7) is known as the **log-likelihood function**.

Now if we differentiate (3.7) partially with respect to $B_1, B_2,$ and σ^2, we obtain

$$\frac{\partial \ln lF}{\partial B_1} = -\frac{1}{\sigma^2}\Sigma(Y_i - B_1 - B_2 X_i)(-1) \tag{3.8}$$

$$\frac{\partial \ln lF}{\partial B_2} = -\frac{1}{\sigma^2}\Sigma(Y_i - B_1 - B_2 X_i)(-X_i) \tag{3.9}$$

$$\frac{\partial \ln lF}{\partial \sigma^2} = -\frac{n}{2\sigma^2} + \frac{1}{2\sigma^4}\Sigma(Y_i - B_1 - B_2 X_i)^2 \tag{3.10}$$

Setting these equations equal to zero (the first-order condition for optimization) and letting $\tilde{b}_1, \tilde{b}_2,$ and $\tilde{\sigma}^2$ denote the ML estimators, we obtain[5]

$$\frac{1}{\tilde{\sigma}^2}\Sigma(Y_i - \tilde{b}_1 - \tilde{b}_2 X_i) = 0 \tag{3.11}$$

$$\frac{1}{\tilde{\sigma}^2}\Sigma(Y_i - \tilde{b}_1 - \tilde{b}_2 X_i)(X_i) = 0 \tag{3.12}$$

$$-\frac{n}{2\tilde{\sigma}^2} + \frac{1}{2\tilde{\sigma}^4}(Y_i - \tilde{b}_1 - \tilde{b}_2 X_i)^2 = 0 \tag{3.13}$$

After simplifying Equations (3.11) and (3.12), we obtain

$$\Sigma Y_i = n\tilde{b}_1 + \tilde{b}_2\Sigma X_i \tag{3.14}$$

$$\Sigma Y_i X_i = \tilde{b}_1\Sigma X_i + \tilde{b}_2\Sigma X_i^2 \tag{3.15}$$

---

[4]Since a log function is a monotonic function, lF, the natural log of LF, will attain its maximum value at the same point as LF.

[5]We use ~ (tilde) for ML estimators to distinguish them from the OLS estimators.

which are precisely the **normal equations** of least-squares theory. Therefore, the ML and OLS estimators of the parameters B_1 and B_2 are the same. This should not be surprising. Examining the log likelihood in Equation (3.7), we observe that the last term in it enters with a negative sign. Therefore, maximizing Equation (3.7) amounts to minimizing this last term, which is precisely the least-squares approach.

If we substitute the ML (=OLS) estimators of the regression parameters into Equation (3.13), after simplification, we obtain the ML estimators of $\tilde{\sigma}^2$ as

$$\tilde{\sigma}^2 = \frac{1}{n}\Sigma(Y_i - \tilde{b}_1 - \tilde{b}_2 X_i)^2$$

$$= \frac{1}{n}\Sigma(Y_i - b_1 - b_2 X_i)^2 \qquad (3.16)$$

$$= \frac{1}{n}\Sigma e_i^2 = \frac{1}{n}e'e$$

The ML estimator of the error variance therefore differs from the OLS estimator, which is $S^2 = \frac{1}{n-2}\Sigma e_i^2$. We have already shown that this is an unbiased estimator of the true error variance, which means that the ML estimator of the error variance is biased.

We can easily determine the magnitude of the bias. Taking the mathematical expectation of Equation (3.16), we obtain

$$E(\tilde{\sigma}^2) = \frac{1}{n}E(\Sigma e_i^2)$$

$$= \frac{1}{n}(n-2)\sigma^2 \qquad (3.17)$$

$$= \sigma^2 - \frac{2}{n}\sigma^2$$

This shows that $\tilde{\sigma}^2$ is biased downward, that is, it underestimates the true error variance. However, as the sample size n increases indefinitely, the bias tends to zero. Therefore, *asymptotically* $\tilde{\sigma}^2$ is unbiased too, that is, $\lim E(\tilde{\sigma}^2) = \sigma^2$ as $n \to \infty$. Furthermore, it can be shown that $\tilde{\sigma}^2$ is also a **consistent estimator** of σ^2, that is, as n increases indefinitely, the ML estimator of the error variance converges to its true value. (Recall the distinction between unbiased and consistency properties of estimators. See Appendix C.)

To further explain the mechanics of the ML method, we consider the **binomial variable** x, which takes the value of zero or one according to the PDF:

$$f(x) = p^x (1-p)^{(1-x)} \quad 0 \le p \le 1; x = 0,1 \tag{3.18}$$

where p is the probability of success and $(1-p)$ is the probability of failure, that is, $f(1) = p$ and $f(0) = 1 - p$. For this model, the likelihood function for a sample drawn at random of size n is

$$LF = \binom{n}{x} = p^x (1-p)^{(n-x)}, \quad x = 0,1,\ldots,n \tag{3.19}$$

Taking the (natural) log of this function, we obtain

$$lF(p) = l \binom{n}{x} + xlp + (n-x)l(1-p) \tag{3.20}$$

where l stands for the natural log (i.e., log to the base e).

Differentiating (3.20) with respect to p and setting the result to zero, we obtain

$$\frac{dlF(p)}{dp} = \frac{x}{p} - \frac{n-x}{1-p} = 0 \tag{3.21}$$

Then we find

$$\hat{p} = \frac{x}{n} \tag{3.22}$$

If we take the second derivative of (3.20), or the first derivative of (3.21), we will find that

$$\frac{d^2 lF(p)}{dp^2} = -\frac{x}{p^2} - \frac{n-x}{(1-p)^2} < 0 \tag{3.23}$$

which shows that $\hat{p}$, the proportion of successes in the sample, does provide the ML estimator of p. (Recall the first-order and second-order conditions for optimization.)

3.3 The Likelihood Function of the k-Variable Regression Model

The counterpart of Equation (3.7) for the k-variable LRM is

$$lF = -\frac{n}{2}\ln(2\pi) - \frac{n}{2}\ln(\sigma^2) - \left[\frac{(y-XB)'(y-XB)}{2\sigma^2}\right] \tag{3.24}$$

If you differentiate this function with respect to B and σ^2, and equate the resulting expressions to zero, you will obtain the ML estimators of the k-regression parameters, which are identical to the OLS estimators given in Equation (1.16), and obtain an estimator of the error variance given in Equation (3.16). Again, note that the ML estimator of the error variance is biased, although the bias will decrease as the sample size increases indefinitely. In practice, however, we use the unbiased estimator of the error variance given in Equation (2.18).

3.4 Properties of the ML Method

We have shown that under the assumptions of the CLRM, the OLS estimators of the regression parameters are BLUE. In establishing this property, we did not make any assumptions about the sampling distribution of the error term, u.

If we use the method of ML, we assume that the error term assumes a particular probability distribution, be it normal, Poisson, logistic, or any other distribution. In regression analysis, it is generally assumed that the error term follows the normal distribution for a variety of reasons discussed previously.

As remarked earlier, the ML estimators are not necessarily unbiased. However, in large samples, they possess several desirable statistical properties. Besides, ML is useful not only in LRMs but also in nonlinear (in-the-parameter) regression models, such as the logit, probit, multinomial, censored, and truncated regression models, and ordinal regression models. We will not deal with nonlinear regression models in this book, for that will require a book by itself.

We now discuss some of the properties of ML estimators. But it should be noted that the majority of the ML properties are *large-sample*, or *asymptotic* ones. Also, the proofs of some of the properties discussed below are technical and are discussed in some detail in Appendix C. The following discussion is therefore heuristic. For advanced discussion of these topics, consult the references.[6]

3.4.1 Consistency of the ML Estimators

Our discussion about the consistency of the OLS estimators carries over to ML estimators. As the sample size gets large enough, estimates obtained

---

[6]Mittelhammer, R. C. (1996). *Mathematical statistic for economics and business*. New York, NY: Springer; Judge, G. G., Carter Hill, R., Griffiths, W. E., Lutkepohl, H., & Lee, T.-C. (1982). *Introduction to the theory and practice of econometrics* (2nd ed.). New York, NY: Wiley; Stachurski, J. (2016). *A primer on econometric theory*. Cambridge, MA: MIT Press.

by the method of ML will be close to the true parameter values with high probability. More technically, if $\tilde{\theta}_n$ is an estimator of θ based on a sample size of n observations, then $\tilde{\theta}_n$ is a consistent estimator of θ if

$$p\lim_{n\to\infty}(\tilde{\theta}_n) = \theta \tag{3.25}$$

where $p$lim means probability limit (see Appendix C).

Consistency is a desirable property in its own right, but it has an important corollary: By Slutsky's theorem, convergence in probability of $\tilde{\theta}$ to θ implies the convergence of any continuous function $f(\tilde{\theta})$ to $f(\theta)$.[7] (For various modes of convergence, see Appendix B. And for a general discussion of the consistency property, see Appendix C.)

3.4.2 Asymptotic Unbiasedness of the ML Estimators

An estimator $\tilde{\theta}$ is said to be an asymptotically unbiased estimator of θ if

$$\lim_{n\to\infty} E(\tilde{\theta}) = \theta \tag{3.26}$$

In words, $\tilde{\theta}$ is an asymptotically unbiased estimator if its expected, or mean, value approaches the true value as the sample size gets larger and larger.

As an example, consider the following sample variance of the variable X.

$$S^2 = \frac{\Sigma(X_i - \bar{X})^2}{n}$$

The usual sample variance has $(n - 1)$ in the denominator because we lose 1 df for using the same sample to calculate the sample mean. It can be shown that

$$E(S^2) = \sigma^2\left(1 - \frac{1}{n}\right)$$

where σ^2 is the true variance. It is obvious that in small sample S^2 is biased, but as the sample size increases indefinitely, $E(S^2)$ approaches the true σ^2, hence S^2 is asymptotically unbiased.

---

[7]On this, see Kramer, J. S. (1986). *Econometric applications of maximum likelihood methods* (chap. 2). New York, NY: Cambridge University Press. Slutsky's theorem states that if Z converges in probability to W, then any continuous function of Z, $f(Z)$, converges to $f(W)$.

3.4.3 Invariance of the ML Estimators

If $\tilde{\theta}$ is an ML estimator of θ and if $f(\theta)$ is a continuous function of θ, then $f(\tilde{\theta})$ is the ML estimator of $f(\theta)$. For example, if $\overline{X}$ is the ML estimator of μ_X, the population mean of the random variable X, then $1/\overline{X}$ is the ML estimator of $1/\mu_X$, provided that $\mu_X \neq 0$. This property does not hold true of the OLS estimators.

As another example, for the LRM $Y_i = B_1 + B_2 X_i + u_i$, the ML estimator of the error variance is given by $\tilde{\sigma}^2 = \dfrac{\Sigma e_i^2}{n}$. Then by the invariance property, the ML estimator of $\tilde{\sigma} = \sqrt{\dfrac{\Sigma e_i^2}{n}}$. That is, the ML estimator of $\tilde{\sigma}$ is the square root of the ML estimator of $\tilde{\sigma}^2$. As yet another example, let $X_1, X_2, \ldots, X_n \sim N(\theta, 1)$. The ML estimator of θ is $\tilde{\theta}_n = \overline{X}_n$. If we consider $\lambda = e^{\theta}$, then the ML estimator of $\tilde{\lambda} = e^{\tilde{\theta}} = e^{\overline{X}}$. The technical details can be found in the references.[8]

3.4.4 Asymptotic Normal Distribution of the ML Estimators

Let $\tilde{\theta}$ be an estimator of the parameter θ. In large samples, technically infinite sample size, it can be shown that

$$\tilde{\theta}^{\text{asy}} \sim N[\theta, I^{-1}(\theta)] \tag{3.27}$$

where $I(\theta)$ is the **information matrix** (see Appendix C) and "asy" means asymptotically. What Equation (3.27) states is that the asymptotic distribution of $\tilde{\theta}$ is normal with mean θ and the variance given by the inverse of $I(\theta)$, where $I(\theta)$ is the information matrix (see Appendix C). It is defined as

$$I(\theta) = -E\left[\frac{\partial^2 lF}{\partial\theta\partial\theta'}\right] \tag{3.28}$$

where the expression in the bracket is the second derivative of the log-likelihood function, which is also known as the **Hessian matrix**, and is discussed in Appendix C.

---

[8]See, for example, Casella, G., & Berger, R. L. (2002). *Statistical inference* (2nd ed., p. 320). Boston, MA: Cengage Learning.

3.4.5 Asymptotic Efficiency of the ML Estimators

Let $\tilde{\theta}$ be an estimator of θ The variance of the asymptotic distribution of $\tilde{\theta}$ is called the **asymptotic variance** of $\tilde{\theta}$. If $\tilde{\theta}$ is consistent and its asymptotic variance is smaller than the asymptotic variance of all other consistent estimators of θ, then $\tilde{\theta}$ is called **asymptotically efficient**. The **Cramer–Rao (CR)** inequality is often used to find out whether an estimator is asymptotically efficient. The CR inequality is generally expressed as

$$\text{var}(\tilde{\theta}) \geq - I^{-1}$$

$$\geq - \frac{1}{E\left[\dfrac{\partial^2 lF}{\partial\theta\partial\theta'}\right]} \tag{3.29}$$

where I is the information matrix given in (3.28) and lF is the log-likelihood function.

The right-hand side of Equation (3.29) is called the **Cramer–Rao lower bound (CRLB)**. The CRLB aids us in constructing a lower limit (greater than zero) for the variance of any unbiased or consistent estimator once we know the functional form of the parent distribution. If we can find an unbiased or consistent estimator whose variance equals the CRLB, then it is the most efficient estimator—we then cannot find another unbiased or consistent estimator with a lower variance.

In Appendix 3A, we discuss the mechanics of the CRLB and show that the ML estimators of the LRM do indeed achieve the CRLB, which is to say that the ML estimators of the LRM are indeed asymptotically efficient.

3.4.6 Sufficiency of the ML Estimators

The concept of sufficient estimator is discussed in Appendix C. Briefly, we say that an estimator or statistic, say, $\tilde{\theta}$, is a sufficient estimator for θ if it *encapsulates* all the information about θ, that is, if it condenses the sample data in such a way that no information about θ is lost. If such a statistic exists, there is no need to examine the entire sample or another statistic based on this sample.

More formally, let $X_1, X_2, \ldots, X_n$ be a random sample from a probability distribution with unknown parameter θ Then the statistic $\tilde{\theta} = f(X_1, X_2, \ldots, X_n)$ is said to be a sufficient statistic for θ if the conditional distribution of $X_1, X_2, \ldots, X_n$, given $\tilde{\theta}$, does not depend on θ

Example:
Consider the Bernoulli probability distribution function

$$f(x;\mathbf{p})=\mathbf{p}^x(1-\mathbf{p})^{1-x} \tag{3.30}$$

where p is the probability of success and $(1-p)$ is the probability of failure in the flip of a coin. Since the coin flips are independent, the likelihood function is the product of n terms like (3.30):

$$\mathbf{LF}=\prod f(X_i;\mathbf{p})=\prod \mathbf{p}^{X_i}(1-\mathbf{p})^{1-X}=\mathbf{p}^s(1-\mathbf{p})^{(n-s)} \tag{3.31}$$

where $s=\Sigma X_i$ and $\prod$ is the product operator.
Therefore the log-likelihood function is

$$lf=s\ln\mathbf{p}+(n-s)\ln(1-\mathbf{p}) \tag{3.32}$$

where ln is the natural log.

Taking the derivative of lf with respect to p and setting it to zero, we get

$$\hat{\mathbf{p}}=\frac{s}{n}=\frac{\Sigma X_i}{n} \tag{3.33}$$

which is simply the average number of successes. But what is important to note is that this estimator of p does not involve the unknown parameter p and is therefore a sufficient statistic for p.

For our purpose, it can be shown that $b=(X'X)^{-1}X'y$ and $S^2=e'e/(n-k)$ are jointly sufficient for B and σ^2. The proof of this statement is rather involved.[9]

3.5 Summary

As an alternative to OLS, we can use the method of ML to estimate the parameters of the classical LRM. But ML requires an explicit assumption about the probability distribution of the error term. In the linear regression analysis, it is usually assumed that the error term follows the normal distribution with zero mean and constant variance. The justification for this is the CLT.

---

[9]For proof, see Myers, H., & Milton, J. S. (1991). *A first course in the theory of linear statistical models* (p. 101). Boston, MA: PWS-Kent.

Although both OLS and ML yield identical estimates of the **B** vector (under the normally distributed error terms), the OLS estimate of the error variance is unbiased but that of the ML estimator is biased. However, this bias diminishes as the sample size increases. The ML estimators possess stronger statistical properties than the OLS estimators, such as the following:

1. ML estimators are invariant to one-to-one transformation, that is, if $\tilde{b}$ is the ML estimator of **B** and if $C = f(B)$ is a function of **B**, then $\tilde{C} = f(\tilde{b})$ is the ML estimator of **C**.

2. They are consistent.

3. They are asymptotically efficient, that is, no asymptotically unbiased estimator has a smaller asymptotic variance.

4. They are asymptotically normally distributed.

5. If there is a sufficient statistic for a parameter, then the ML estimator of the parameter is a function of the sufficient statistic.

6. They are uniformly minimum variance unbiased estimators (UMVUE).

However, the mathematics of ML estimation is not trivial, especially if confidence intervals for parameters are desired. Also, the likelihood function needs to be carefully worked out for a given (probability) distribution. The properties of ML discussed earlier generally do not hold in small samples.

Exercises

3.1 Consider the binomial variable x discussed in the chapter.
   a. If a random sample of n observations is drawn from this PDF, what is the ML estimator of p and the variance of its sampling distribution?
   b. What is the asymptotic variance of the ML estimator found in (b)?
   c. *Does $\hat{p}$, the estimator of p, attain the CRLB for the variance of an unbiased estimator? (*Optional)

3.2 A random variable x is said to follow the Poisson distribution if it has the following density function:

$$f(x;\lambda) = \frac{\lambda^x e^{-\lambda}}{x!}, \lambda > 0, x = 0, 1, 2, \ldots$$

where λ is the parameter of the distribution. A sample of size n is drawn randomly from this distribution.

    a. Obtain the likelihood function for the sample.

    b. Estimate the parameter λ. How would you interpret it?

    c. What is the variance of this distribution?

    d. Does it attain the CRLB?

3.3 $X_1, X_2, ..., X_n$ is a random sample from $N(\lambda, 1)$.

    a. What is the ML estimator of λ?

    b. Let $\gamma = e^\lambda$. What is the ML estimator of γ? (*Hint:* Invariance property of ML estimator.)

Appendix 3A: Asymptotic Efficiency of the ML Estimators of the LRM

We start with[10]

$$lF = -\frac{n}{2}\ln(2\pi) - \frac{n}{2}\ln(\sigma^2) - \left[\frac{(y - XB)'(y - XB)}{2\sigma^2}\right] \tag{3A.1}$$

From this log-likelihood function, we obtain the following function:

$$W(B, \sigma) = \begin{bmatrix} \dfrac{\partial lF}{\partial B} \\ \dfrac{\partial lF}{\partial \sigma} \end{bmatrix} = \begin{bmatrix} \dfrac{X'(y - XB)}{\sigma^2} \\ -\dfrac{n}{\sigma} + \dfrac{1}{\sigma^3}(y - XB)'(y - XB) \end{bmatrix} \tag{3A.2}$$

W is called the **score function**, which is simply the partial first derivative of the log-likelihood function (see Appendix C).

To obtain ML estimators of the parameters, we set the score function to **0**, which gives

$$X'(y - XB) = 0 \tag{3A.3}$$

$$\frac{(y - XB)'(y - XB)}{\sigma^3} = \frac{n}{\sigma} \tag{3A.4}$$

---

[10]See Ramanathan, R. (1993). *Statistical methods in econometrics* (pp. 191–192). New York, NY: Academic Press.

These equations yield

$$\tilde{b} = (X'X)^{-1}X'y \qquad (3A.5)$$

$$\tilde{\sigma}^2 = \frac{1}{n}(y - X\tilde{b})'(y - b\tilde{X}) = \frac{1}{n}(e'e) \qquad (3A.6)$$

As we noted previously, the OLS and ML estimators of b are identical, but their error variances are different: The OLS error variance is unbiased, but the ML variance is biased in finite samples.

Now to find out if the ML error variance attains the CRLB, we need to estimate the **information matrix**, also called the **Fisher information matrix**—the amount of information a sample provides about the value(s) of the unknown parameter(s).

It is easy to verify that

$$\frac{\partial^2 lF}{\partial B^2} = -\frac{X'X}{\sigma^2} \qquad (3A.7)$$

$$\frac{\partial^2 lF}{\partial \sigma^2} = \frac{n}{\sigma^2} - \frac{3}{\sigma^4}(y - XB)'(y - XB) \qquad (3A.8)$$

Furthermore,

$$E\left[-\frac{\partial^2 lF}{\partial^2 B^2}\right] = \frac{X'X}{\sigma^2} \qquad (3A.9)$$

$$E\left[-\frac{\partial^2 lF}{\partial^2 \sigma^2}\right] = -\frac{n}{\sigma^2} + \frac{3}{\sigma^4}E[(y - XB)'(y - XB)]$$

$$= -\frac{n}{\sigma^2} + \frac{3}{\sigma^4}E(u'u)$$

$$= -\frac{n}{\sigma^2} + \frac{3}{\sigma^4}(n\sigma^2) \qquad (3A.10)$$

$$= -\frac{n}{\sigma^2} + \frac{3n}{\sigma^2}$$

$$= \frac{2n}{\sigma^2}$$

$$\frac{\partial^2 lF}{\partial \sigma \partial B} = -\frac{2}{\sigma^3} X'(y - XB) \tag{3A.11}$$

Since $E(y|X) = XB$,

$$E\left[-\frac{\partial^2 lF}{\partial \sigma \partial B}\right] = 0 \tag{3A.12}$$

Collecting these results, the information matrix $I(\theta)$ and its inverse are as follows:

$$I(\theta) = \begin{pmatrix} \dfrac{X'X}{\sigma^2} & 0 \\ 0 & \dfrac{2n}{\sigma^2} \end{pmatrix} \tag{3A.13}$$

$$[I(\theta)]^{-1} = \begin{pmatrix} \sigma^2(X'X)^{-1} & 0 \\ 0 & \dfrac{\sigma^2}{2n} \end{pmatrix} \tag{3A.14}$$

Since these matrices are block-diagonal, we can treat B and σ^2 independently for the purpose of statistical inference. We have already found that B and σ^2 are uncorrelated, but with the normality assumption for u, they are also independent.

Note: In deriving these results, we have used $y - XB = u$ and note that $E(u_i) = 0$. As you can see from Equation (3A.14), the CRLB for the variance–covariance of b (the estimator of B) is precisely the same as the one we obtained from the ML method. This is to say, the variance–covariance matrix of the ML estimators does indeed attain the CRLB. To put it yet another way, the ML estimators of the LRM are the most efficient (compare a similar statement we made about the Gauss–Markov theorem).

CHAPTER 4. LINEAR REGRESSION MODEL: DISTRIBUTION THEORY AND HYPOTHESIS TESTING

4.1 Introduction

In Chapters 2 and 3, we discussed at length the methods of OLS and ML to estimate the parameters of the LRM. We also discussed the small- and large-sample properties of these estimators. Our task in this chapter is to draw inferences about the "true" parameter values of the LRM. Since we rarely observe the LRM directly, and since we usually do not have access to the whole population underlying the LRM, we have to base our analysis on the basis of a randomly drawn sample from the purported population.

On the basis of the sample at hand, we obtain estimates of the population parameters using OLS or ML estimating procedures. As we saw, for all practical purposes, the OLS and ML estimators are identical. But in LRM, the ML method explicitly assumes that the error term is normally distributed with zero mean and constant variance. As we noted in Chapter 2, for statistical inference, we need to assume some probability distribution for the error term. And for reasons already discussed, we assume that the error term is normally distributed.

Since OLS and ML estimators are based on a sample, the values of the estimators will differ from sample to sample. The sampling variability of the estimators can be judged by their variances or their square roots, **the sample standard errors**. For the purpose of statistical inference, we need an appropriate **test statistic** to judge whether an observed value of an estimator can reasonably come from the hypothesized value of the estimator. In the remainder of this chapter, we will discuss several test statistics.

4.2 Types of Hypotheses

Remember that the generic LRM is as shown in Equation (1.2), which for convenience is reproduced below:

$$Y_i = B_1 + B_2 X_{2i} + B_3 X_{3i} + \cdots + B_k X_{ki} + u_i \qquad (4.1)$$

The corresponding sample regression is

$$Y_i = b_1 + b_2 X_{2i} + b_3 X_{3i} + \cdots + b_k X_{ki} + e_i \qquad (4.2)$$

Once we have a concrete sample, we can obtain values of the intercept and the partial regression coefficients, using the formula

$$b = (X'X)^{-1}X'y \quad (1.16)$$

and obtain the variances of the regressing coefficients using

$$\text{cov}(b) = \sigma^2 (X'X)^{-1} \quad (2.8)$$

and obtain the estimator of σ^2 as

$$S^2 = e'e/(n-k) \quad (2.18)$$

Of course, all these estimates are obtained from a statistical software package.

Note that $\text{cov}(b)$ is often used as a short form for variance–covariance matrix, the entries on the main diagonal giving the variances and the off-diagonal entries giving the covariances between pairs of regression coefficients.

4.3 Procedure for Hypothesis Testing

First, the term *hypothesis* means a conjecture about the value that a parameter might take. For example, in the LRM (4.1), we might state that the true value of the parameter B_2 is, say, 0.75. Given the sample data, we obtain the value of b_2, say, 0.69. Is 0.69 *statistically different* from the hypothesized value of 0.75? How do we decide that? We obviously need a procedure to resolve this question. The hypothesis-testing procedure involves three steps: (1) formulate what is called a **null hypothesis** and an **alternative hypothesis**, (2) develop a **test statistic** and its **sampling or probability distribution**, and (3) decide a **decision rule** regarding the null and alternative hypotheses.

4.3.1 Null and Alternative Hypotheses

The term *null hypothesis* was introduced by the statistician Sir Ronald Fisher in 1935. It is also called the **maintained hypothesis**, the hypothesis we want to test. In statistics, it is usually denoted by H_0, and it is tested against an *alternative hypothesis*, denoted by H_1. The alternative hypothesis can be **composite** or **one-sided**. In the hypothetical case cited earlier, the null hypothesis is $H_0 = 0.75$ and the alternative hypothesis could be $H_1 \neq 0.75$ (a composite hypothesis), meaning B_2 could be less than 0.75 or greater than 0.75, or it could be $H_1 > 0$ or $H_1 < 0$, which are one-sided hypotheses.

4.3.2 Test Statistic

To test the null hypothesis, we use the sample information to obtain what is known as a **test statistic**. Very often, the test statistic is based on the

point estimator of the parameter, such as b_2 as an estimator of B_2 in a regression model. Then, we try to find the **sampling**, or **probability distribution** of the test statistic, and use the **confidence interval**, or a **test of significance approach**, to test the null hypothesis. We will illustrate these approaches with the example of the wage regression introduced in Chapter 1.

Typically, the null hypothesis that is chosen is that a particular B coefficient is zero or a group of B coefficients is zero. The idea is to find out if a particular regressor or a group of regressors have any impact on the regressand to begin with before one can test more complicated hypotheses.

The test statistics that are frequently used in regression analysis are the Z **test**, the t **test**, the **chi-square** (χ^2) **test**, and the F **test**. The probability distributions of these tests are discussed in Appendix D. The reader should read this appendix to learn about the features of each of these distributions.

4.3.3 Decision Rule

On the basis of the chosen test statistic, we either reject the null hypothesis or do not reject it. Since the test statistic is a random variable, in principle it can take any value. To decide which of the values are compatible with the null hypothesis and which are not, we need to decide in advance on a **rejection rule**, called the **critical region**, so that if the value of the test statistic falls in the **rejection region**, we reject the null hypothesis, but if it does not fall in that region, we do not reject the null hypothesis.

Now it is possible that the chosen test rejects the null hypothesis that is true, in which case we make what is called a **Type I error**. On the other hand, if we do not reject the null hypothesis even if it is not true, we commit what is known as a **Type II error**. Put differently, rejecting H_0 when H_0 is true is a Type I error and retaining H_0 when H_1 is true is a Type II error.

The probability of making a Type I error is called the **level of significance** of the test and is usually denoted by α. In practice, the α value is generally fixed at the 0.10 (10%), 0.05 (5%), or 0.01 (1%) levels. The 5% level of significance means that if the null hypothesis is true, we will not reject it more than 5% of the time. If $\alpha = 1\%$, it means we do not reject the null hypothesis more than 1% of the time. The probability of making a Type II error is denoted by β. Schematically,

	Reject H_0	Do not reject H_0
H_0 is true	Type I error	Correct decision
H_0 is false	Correct decision	Type II error

Ideally, we would like to minimize both Type I and Type II errors. Unfortunately, for any given sample size, it is not possible to minimize both of the errors simultaneously. The classical approach to this problem, embodied in the work of **Neyman** and **Pearson**, is to assume that a Type I error is likely to be more serious in practice than a Type II error: In court cases, it is generally assumed that a person is innocent unless the evidence strongly suggests that he or she is guilty. In statistics, we retain H_0 unless there is strong evidence to reject H_0. Therefore, one should try to keep the probability of committing a Type I error, α, at a fairly low level, such as 1% or 5%, and then try to minimize the probability of a Type II error, β, as much as possible.

When testing the null hypothesis that the unknown parameter $\theta = \theta_0$ against the alternative hypothesis $\theta = \theta_1$, the quantity $1 - \beta$ is called the **power of the test** at $\theta = \theta_1$. In other words, the power of the test is the probability of not rejecting the alternative hypothesis when it is in fact true and its value depends on the value of the particular alternative and on the sample size. Put differently, the power of a test is its ability to reject a false null hypothesis. A graph of $1 - \beta$ as a function of the true value of the parameter is called the **power curve** for the statistical test. Ideally, one would like α to be small and the power $1 - \beta$ to be as large as possible.

A critical region for testing a simple null hypothesis $\theta = \theta_0$ against a simple alternative hypothesis $\theta = \theta_1$ is called a **best critical region** or a **most powerful critical region**, if the power of the test at $\theta = \theta_1$ is a maximum.

It is important to note that failure to reject the null hypothesis does not mean that it is true. It may mean that the test does not have power against the particular true hypothesis.

4.3.3.1 The p Value

Instead of preselecting α at arbitrary levels, such as 1%, 5%, or 10% for deciding whether to reject or not reject a null hypothesis, a more sophisticated approach is to calculate the **p (probability) value**, the **exact level of significance**, or the **marginal level of significance** of the test statistic. It is defined as the *lowest significance level at which a null hypothesis can be rejected*. Put differently, the null hypothesis should be rejected for all small values of p and should not be rejected for large values of p. In other words, the lower the p value, the greater the evidence against the null hypothesis. Note that the p value depends on whether we are considering a one-sided or two-sided (i.e., composite) alternative hypothesis.[1]

Before we discuss various hypotheses, let us revisit the wage regression given in Equation (1.40). We will use this example to shed light on various hypotheses.

---

[1]On the abuse and misuse of p values, see Trafimow, D., & Marks, M. (2015). Editorial. *Basic and Applied Social Psychology, 37*(1), 1–2.

4.4 The Determination of
Hourly Wages in the United States

In Section 1.6, in Equation (1.40) we provided an example of a wage regression that was based on the U.S. data for 1995. The details of the data and the variables used in the analysis are also given in Section 1.6. The wage model we used is

$$W_i = B_1 + B_2 FE_i + B_3 NW_i + B_4 UN_i + B_5 ED_i + B_6 EX_i + u_i \qquad (4.3)$$

These variables are as defined in Section 1.6.

A priori, we would expect *Education* and *Experience* to have a positive impact on hourly wages. In relation to nonunion workers, hourly wages are expected to be higher for union workers. Compared with male workers, female workers, on average, are expected to have lower hourly wages. Likewise, compared with white workers, nonwhite workers are expected to have lower hourly wages. These expectations are based on historical labor market experience.

The full results underlying Equation (1.40) based on OLS are given in Table 4.1 and are obtained from the *Eviews 9* statistical package.

Table 4.1 is divided into three parts. The *first part* gives the name of the dependent variable, or the regressand; the number of observations in the sample; and the method used in the analysis, OLS in the present instance. The *second part* of the table gives the regressors used in the model, the point estimates of their coefficients, their estimated standard errors, and the *t* statistics and their probability or *p* values. *All the* t *values presented in this table are obtained under the null hypothesis that the true value of each parameter is zero.* The *C* value in this table represents the intercept. The *third part* of the table gives some descriptive statistics, such as R^2, adjusted R^2, standard error of the regression (i.e., *S*), value of the RSS, and several other statistics, which we will discuss at the appropriate time. But note the value of -4240.37 of the log-likelihood function. We will discuss this value later in the chapter.

4.4.1 Interpretation of the Estimated Regression

The negative intercept value usually has no viable economic meaning. Literally interpreted, it means if the values of all the regressors are set to zero, the average wage would be about $\$-7.18$, which obviously makes no sense. However, there are some special cases where the negative value of the intercept is meaningful. The *Union* coefficient 1.09 means, holding other regressor values constant, that union workers on average make about

Table 4.1 The Wage Regression

Dependent Variable: W
Method: Least Squares

Sample: 1 1289
Included observations: 1289

Variable	Coefficient	Std. Error	t-Statistic	Prob.
C	-7.183338	1.015788	-7.071691	0.0000
FE	-3.074875	0.364616	-8.433184	0.0000
NW	-1.565313	0.509188	-3.074139	0.0022
UN	1.095976	0.506078	2.165626	0.0305
ED	1.370301	0.065904	20.79231	0.0000
EX	0.166607	0.016048	10.38205	0.0000

R-squared	0.323339	Mean dependent var	12.36585
Adjusted R-squared	0.320702	S.D. dependent var	7.896350
S.E. of regression	6.508137	Akaike info criterion	6.588627
Sum squared resid	54342.54	Schwarz criterion	6.612653
Log likelihood	-4240.370	Hannan-Quinn criter.	6.597646
F-statistic	122.6149	Durbin-Watson stat	1.897513
Prob(F-statistic)	0.000000		

$1.09 per hour more than the nonunion workers. The *Female* coefficient of -3.07 means that the average hourly wage of female workers is about $3.07 lower than their male counterparts, again holding the other regressor values constant. The average hourly wage of nonwhite workers is about $1.56 lower than white workers, ceteris paribus. If the years of education increases by a year, the average hourly wage goes up by about $1.37, ceteris paribus. If potential work experience goes up by a year, the average hourly wage goes up by about $0.17. All these results accord with prior expectations about labor market behavior in the United States. The R^2 value of 0.32 suggests that about 32% of the variation in hourly wages is explained by the regressors included in the model. The adjusted-R^2 value is slightly lower, as per theory.

4.5 Testing Hypotheses About an Individual Regression Coefficient

Suppose in the model (4.1), we want to test the following hypotheses about the regression coefficient B_k:

$$H_0: B_k = 0$$
$$H_1: B_k \neq 0 \tag{4.4}$$

The alternative hypothesis is two-sided, meaning B_k could be greater than zero or less than zero. The estimated value of B_k is given by b_k.

Now from Equation (2.9a), we know that

$$b_k \sim N(B_k, \sigma^2 x^{kk}) \tag{4.5}$$

Now consider the following statistic:

$$Z = \frac{b_k - B_k}{se(b_k)} = \frac{b_k - B_k}{\sigma\sqrt{x^{kk}}} = \frac{b_k}{\sigma\sqrt{x^{kk}}} \tag{4.6}$$

given that $B_k = 0$ under the null hypothesis, where se stands for standard error.

Since Z in Equation (4.6) is a standardized normal variable, that is, $Z \sim N(0,1)$, we can use the normal distribution to test the null hypothesis given in Equation (4.4). We call it the **Z test**. Since the normal distribution is well-known and the standardized normal distribution tables are so readily available, it should be an easy task to test the null hypothesis. Notice that a computed Z value can be positive or negative.

However, to apply the Z test, we must know σ^2, the variance of the error term u. We do not know it directly, but we estimate it from its unbiased estimator, S^2, given in Equation (2.18). So if we put S^2 for σ^2 (or S for σ) into (4.6), we obtain another statistic, called the **t statistic**.[2]

$$t = \frac{b_k - B_k}{S\sqrt{x^{kk}}} = \frac{b_k}{S\sqrt{x^{kk}}} \quad (\text{if } B_k = 0) \tag{4.7}$$

The t statistic follows the Student's t distribution, which is discussed in Appendix D. Under the null hypothesis $B_k = 0$, the t value is simply the ratio

---

[2]If variable Z_1 is a standard normal variable and another variable Z_2 is a chi-square variable with k df and is distributed independently of Z_1, then $t = Z_1/\sqrt{Z_2/k}$ follows the t distribution with k df. See Appendix D.

of b_k to its standard error. All the t values shown in Table 4.1 are thus the ratios of the estimated coefficients divided by their standard errors, under the null hypothesis that the relevant population coefficients are zero.

The t distribution, like the standard normal distribution, is centered at zero, is symmetrical, but is flatter than the normal distribution, that is, it is more spread out because it has larger variance than the standard normal distribution. As noted in Appendix D, the mean of the t distribution is 0 and its variance is $k/(k-2)$ where k is the degrees of freedom and is defined for $k > 2$. Electronic tables of the t distribution are readily available. A unique feature of the t distribution is that it depends on the **degrees of freedom**, which is the number of *independent* observations available for analysis. If a regression model is estimated with n observations and there are k regression coefficients estimated, then the degrees of freedom are $(n-k)$. Like the Z value, a t value can be positive or negative.

Now let us turn to Table 4.1. Suppose we want to test the null hypothesis that education has no impact on average hourly wages. And assume that the alternative hypothesis is that education does have an impact on average hourly earnings, positive or negative. The t value of the education coefficient of 20.8 has a p value of practically zero, which suggests that we should reject the null hypothesis that education has no impact on average hourly earnings, holding other variables constant. The null hypothesis in this case is obviously false, so carrying out a formal test is not necessary, but we do it just to illustrate the methodology. If the null hypothesis were true, the probability of obtaining a t value of as much as 20.8 is extremely small. Therefore, it is extremely unlikely that the null hypothesis is true. When we obtain a high t value in absolute terms (i.e., disregarding the sign) and a very low p value, we say the results are **highly statistically significant**, meaning we overwhelmingly reject the null hypothesis.

If you look at the other regressors in Table 4.1, you will see that all the regression coefficients are highly significant, as their p values are so low. What this means is that each of the regressors in Table 4.1 affects the average hourly earnings. The term *highly significant* means a very low p value; in practice an α value of 1% is deemed a sufficiently low p value.

We could have rejected the null hypothesis at the conventionally used α values of 10%, 5%, or 1%. But a great advantage of the p value is that it gives us the exact level of significance without needing to choose α values arbitrarily.

Note: The sample size in Table 4.1 of 1,289 is quite large. If you consult the t table in any statistics textbook, you will find that for $\alpha = 5\%$, the t value is about 1.96, which is the same as that for the standard normal distribution. This is not surprising, for as the sample size increases, the t distribution approaches the standard normal distribution.

4.5.1 One-Sided or Two-Sided Alternative Hypotheses

All the p values given in Table 4.1 are for a two-sided alternative hypotheses. But sometimes the alternative hypothesis could be one-sided, such as

$$H_0: B_4 \text{ (coefficient of education)} = 0$$
$$H_1: B_4 > 0$$

or

$$H_1: B_4 < 0$$

For such one-sided alternative hypotheses, the p value is **one-half** the p value computed for the two-sided hypothesis. In our illustrative example, it does not make sense to make $H_1: B_4 < 0$; if that were the case, it would be bad news for education. The more appropriate alternative hypothesis in this case is $H_1: B_4 > 0$.

It should be noted that the t and p values given in the preceding tables are for testing the zero null hypothesis. But if we want to test the hypothesis, say, $B_k = 1$, the t and p values in the output are not relevant for this test.

4.5.2 A Note on Rejecting or Accepting a Null Hypothesis

If a test shows that we cannot reject a null hypothesis, it does not mean that we have positive evidence for the null hypothesis. As Hendry and Nielsen note:[3]

> This is so, since the decision is based on the assumption that the statistical model is valid. . . . An important consequence is that testing can be used for *falsifying* economic theories, but we support an economic theory only insofar as we are not able to falsify it.

4.5.3 Confidence Interval Approach to Hypothesis Testing

Instead of testing a specific hypothesis, we can establish a range or an interval that has a certain probability of including the true parameter value. To be specific, in our wage example, we found that the coefficient of education is about 1.37, which is a point estimate of the true parameter B_4, the education coefficient in the LRM.

To establish a confidence interval, say for B_4, we can use (4.7), as follows:

$$\Pr(-t_{\alpha/2} \le t \le t_{\alpha/2}) = 1 - \alpha \tag{4.8}$$

---

[3]Hendry, D. F., & Nielsen, B. (2007). *Econometric modeling: A likelihood approach* (pp. 24–25). Princeton, NJ: Princeton University Press.

where the t in the middle of this double inequality is given by Equation (4.7), which in the present case is

$$t = \left[\frac{b_4 - B_4}{se(b_4)} \right] \tag{4.9}$$

where α is the level of significance and where $t_{\alpha/2}$ is the value of the t variable obtained from the t distribution for the $\alpha/2$ level of significance and for $(n-k)df$. It is called the **critical** t value at the $\alpha/2$ level of significance. Putting t from Equation (4.9) into Equation (4.8) and rearranging, we obtain

$$\Pr\left[b_4 - t_{\alpha/2}se(b_4) \le B_4 \le b_4 + t_{\alpha/2}se(b_4)\right] = 1-\alpha \tag{4.10}$$

Equation (4.10) provides a $100(1-\alpha)\%$ confidence interval for B_4, which can be written compactly as (see Figure 4.1)

$$\Pr(b_4 \pm t_{\alpha/2}se(b_4)) = (1-\alpha) \tag{4.11}$$

The smaller this interval is, the more confidence we have in the estimated value of the parameter. It is critical that we know how to interpret the interval (4.10). First, note that this interval is random because the values of the estimated coefficient and its standard error will vary from sample to sample, although B_4 is fixed. Therefore, what Equation (4.10) states is that this random interval has probability $(1-\alpha)$ of containing the true but unknown parameter B_4. It is not correct to say that the probability is 95% that B_4 lies in this interval; it is the interval that is random and not B_4.

Figure 4.1 Confidence Interval for B_4

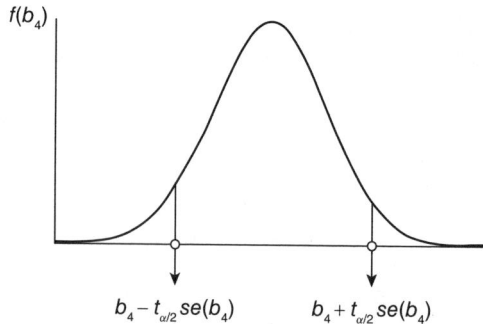

$$b_4 - t_{\alpha/2}se(b_4) \qquad b_4 + t_{\alpha/2}se(b_4)$$

To use Equation (4.11), we know the values of b_4 and its standard error, but we need to choose the value of α and the associated t value, which depends on the degrees of freedom. If the sample size is reasonably large, we can use the Z value of the standard normal distribution. For example, if $\alpha = 5\%$, the Z value is 1.96. So to establish a 95% confidence interval for B_4, we use $b_4 + 1.96se(b_4)$. From Table 4.1, we obtain this as

$$1.3703 \pm 1.96(0.0659)$$

that is,

$$(1.2411, 1.4995) \tag{4.12}$$

Although the interval in (4.10) is random, the interval (4.12) is *not* random; it is fixed. So either B_4 lies in it or it does not. In general, for any value C contained in the interval, the test H_0: $B_k = C$ would not be rejected using the two-sided test, but for any value outside the interval, the null hypothesis would be rejected.

The idea of an interval estimator applies to other parameters of the LRM. For example, we can obtain an interval estimator for the true error variance, as we will show shortly. But keep in mind that any single interval estimator may not contain the true parameter value. But if we establish such intervals in *repeated sampling*, then on average, such intervals will contain the true parameter value with certain confidence. Incidentally, several statistical packages produce desired confidence intervals for each of the estimated regression coefficients.

4.6 Testing the Hypothesis That All the Regressors Collectively Have No Influence on the Regressand

Although this hypothesis may not be meaningful in the present example, we have to discuss this case as a theoretical possibility. This hypothesis is often called the **overall significance of the regression**. The null and alternative hypotheses here are as follows:
In the LRM (4.1),

$$H_0: B_2 = B_3 = \ldots = B_k = 0$$

H_1: At least one partial regression coefficient is not zero statistically.

If the null hypothesis is true, it would mean the average value of the regressand is equal to the value of the intercept. In our example, this would not make sense, but we are talking generally.

To test the present null hypothesis, we use the **F test**. This test is based on the **F distribution**. Appendix D gives the details of this distribution. Briefly, F is the ratio of two *independently* distributed chi-square variables divided by their respective degrees of freedom. For the present purpose, we define it as follows:

$$F = \frac{\text{Explained sum of squares}/df_1}{\text{Residual sum of squares}/df_2} = \frac{\text{ESS}/df_1}{\text{RSS}/df_2} \tag{4.13}$$

We have already come across these terms in Equation (1.34). The degrees of freedom for $\text{ESS}(df_1)$ is the number of partial regression coefficients estimated, 5 in our wage example. The degrees of freedom for the RSS (df_2) is the number of observations (1,289 in our example) minus the total number of coefficients estimated, which is 6 in our case—intercept and five partial regression coefficients, which amounts to 1,283.

Several statistical packages produce a table, called the **analysis of variance (ANOVA)** table, which breaks down the various sums of squares involved in regression analysis and also presents the F value. *Eviews* does not give the ANOVA table explicitly, although it produces the F statistic, as shown in the third part of Table 4.1.

Using *Stata 13*, we obtain the ANOVA table (Table 4.2).

In Table 4.2, ESS is represented by the model (sum of squares), and RSS is referred to as residual SS. Dividing these sums of squares by their *df*, we obtain MS, which is the mean, or average, sum of squares. Using Equation (4.13), we obtain the F statistic, which is 122.61. This F value for 5, the numerator *df*, and 1,283, the denominator *df*, is highly significant, for the p value of obtaining such an F value, if the null hypothesis were true, is practically zero (see also the F value in Table 4.1). Therefore, we reject the null hypothesis: In our wage regression, at least one regressor significantly affects the average hourly earnings.

Table 4.2 ANOVA for the Wage Regression

Source	SS	df	MS	Number of Obs. = 1289
				$F(5, 1283) = 122.61$
Model	25967.2805	5	5193.45611	Prob > $F = 0.0000$
Residual	54342.5442	1283	42.3558411	R-squared = 0.3233
				Adj R-squared = 0.3207
Total	80309.8247	1288		

4.6.1 Relationship Between the t Test and the F Test

The square of the t variable with k degrees of freedom has an F distribution with $k_1 = 1$ df in the numerator and k df in the denominator. That is,

$$F_{1,k} = t_k^2 \tag{4.14}$$

Note that for this equality to hold, the numerator df of the F variable must be 1.

4.6.2 Relationship Between R² and F Tests of Significance

There is an intimate relationship between R^2 and F. Assuming that the error term in the k-variable regression model (4.1) is normally distributed, to test the hypothesis that all the partial slope coefficients are zero, we use the F test:

$$F = \frac{\text{ESS}/(k-1)}{\text{RSS}/(n-k)} \tag{4.15}$$

which follows the F distribution with $(k-1)$ and $(n-k)df$.

Manipulating (4.15), we obtain

$$\begin{aligned}
F &= \frac{n-k}{k-1} \frac{\text{ESS}}{\text{RSS}} \\
&= \frac{n-k}{k-1} \frac{\text{ESS}}{\text{TSS}-\text{ESS}} \\
&= \frac{n-k}{k-1} \frac{\text{ESS/TSS}}{1-(\text{ESS/TSS})} \\
&= \frac{n-k}{k-1} \frac{R^2}{1-R^2} \\
&= \frac{R^2/(k-1)}{(1-R^2)/(n-k)}
\end{aligned} \tag{4.16}$$

This equation shows how F and R^2 are intimately related. These two vary directly. *Thus the F test, which is a test of the overall significance of the estimated regression, can also be regarded as a test of significance of $R^2 = 0$.*

Since R^2 is routinely computed for the regression model, we can use it to test the overall significance of the estimated regression without having to know the TSS, ESS, and RSS.

For the illustrative example, applying (4.16), we obtain

$$F = \frac{0.3233/5}{(1-0.3233)/(1289-6)} \approx 122.60$$

which is about the same as we obtained in Table 4.2. For 5 and 1,283 df, this F value is highly significant, that is, we reject the null hypothesis that all the regressor coefficients are equal to zero.

4.7 Testing the Incremental Contribution of a Regressor

Suppose in our wage regression, initially we decided not to include the dummy variables, nonwhite and union? Now we raise this question: Should we have included these variables to begin with? The results in Table 4.1 would definitely suggest that we should have done so. But to illustrate the incremental contribution of these regressors, let us estimate the wage model without these two regressors. We call this model the **restricted LRM (RLRM)**, and call the original model with all the variables included, the **unrestricted LRM (UNLRM)**. Sometimes these are known as *short regression* and *long regression*.

The restricted wage regression, therefore, is

$$W_i = B_1 + B_2 FE_i + B_5 ED_i + B_6 EX_i + v_i \qquad (4.17)$$

In this restricted regression, we have replaced the usual error term u_i by v_i to distinguish it from the unrestricted regression given in Equation (4.1), since the restrictions $B_3 = B_4 = 0$ have been imposed.

Table 4.1 gives us the results of the UNLRM. The results of the RLRM are shown in Table 4.3.

If you compare the result of the restricted wage regression (Table 4.3) and the unrestricted wage regression (Table 4.1), we see that the values the regressors included in both the tables are slightly different, but they are all statistically significant. The R^2 value in Table 4.3 is slightly lower than that in Table 4.1. This is not surprising, for R^2 generally increases with the number of regressors in the model. Should we then include the omitted variables or not include them?

We can answer this question with the F test as follows: Let

RSS_R = residual sum of squares from the restricted regression (Table 4.3)

RSS_{UR} = residual sum of squares from the unrestricted regression (Table 4.1)

Table 4.3 Restricted Wage Regression (Union and Nonwhite Variables Dropped)

Dependent Variable: W
Method: Least Squares
Sample: 1 1289
Included observations: 1289

Variable	Coefficient	Std. Error	t-Statistic	Prob.
C	−7.653033	1.005982	−7.607522	0.0000
FE	−3.186426	0.364427	−8.743651	0.0000
ED	1.393727	0.065841	21.16813	0.0000
EX	0.174712	0.015881	11.00115	0.0000

R-squared	0.316487	Mean dependent var		12.36585
Adjusted R-squared	0.314891	S.D. dependent var		7.896350
S.E. of regression	6.535911	Akaike info criterion		6.595599
Sum squared resid	54892.81	Schwarz criterion		6.611616
Log likelihood	−4246.864	Hannan-Quinn criter.		6.601611
F-statistic	198.3312	Durbin-Watson stat		1.902639
Prob(F-statistic)	0.000000			

m = number of linear restrictions (2 in the present example)

k = number of parameters in the unrestricted regression (6 in the present case)

n = number of observations (1,289 in our example)

Now consider the following F statistic:

$$F = \frac{(\text{RSS}_R - \text{RSS}_{UR})/m}{\text{RSS}_{UR}/(n-k)} \sim F_{m,(n-k)} \quad (4.18)$$

It should be noted that RSS_R can never be smaller than RSS_{UR}, so this F ratio is always nonnegative. As usual, if the F computed from Equation (4.18) exceeds the critical F value at the chosen level of significance for the appropriate degrees of freedom, we reject the null hypothesis that the restrictions are valid; otherwise we do not reject it.

For our example,

$$\text{RSS}_R = 54,893 \quad (\text{Table 4.3})$$
$$\text{RSS}_{UR} = 54,342 \quad (\text{Table 4.1})$$

Therefore,

$$F = \frac{54893/2}{54342/1283} = 275.5/42.35 \approx 6.50$$

This significant F suggests that we reject the null hypothesis that the variables nonwhite and union do not belong in the wage regression. To put it simply, we should not be omitting these two variables from the wage regression.

Incidentally, this example illustrates the point made in Chapter 2 that the regression model must be *correctly* specified for the estimates to be unbiased. By omitting the two important variables, we have committed what is called a form of **specification error**, the error of omitting (theoretically) relevant variables. Note that the log-likelihood value for the restricted wage regression is -4246.864. Disregarding the sign, this number is greater than the one for the full model. We will see the utility of this in Section 4.9. We discuss the topic of specification bias in Chapter 7.

This example is a specific case of **regression subject to linear restrictions**. Appendix 4A presents this more generally using matrix algebra. *Eviews*, *Stata*, and several other statistical packages have procedures for testing linear restrictions.

4.8 Confidence Interval for the Error Variance σ^2

To derive a confidence interval for the unknown error variance, σ^2, we know from Equation (1.20) that

$$e = My \quad (1.20)$$

where $M = I - X(X'X)^{-1}X'$.

We showed earlier that M is symmetric and idempotent, and $MX = 0$. Now

$$e = My = M(XB + u) = Mu \quad (4.19)$$

Therefore,

$$e'e = u'Mu \quad (4.20)$$

Since by assumption $u \sim N(0, \sigma^2 I)$, $\dfrac{u}{\sigma} \sim N(0,1)$, that is, it is a standard normal variable. Now

$$\frac{e'e}{\sigma^2} = \frac{u'M'Mu}{\sigma^2} = \frac{u'Mu}{\sigma^2} \sim \chi_r^2 \tag{4.21}$$

where r is the rank of M. As Stewart and Gill note, "A quadratic form in a standard normal vector and a symmetric idempotent matrix has a χ^2 distribution with degrees of freedom equal to the rank of the idempotent matrix."[4]

Since the rank of a symmetric idempotent matrix is equal to its trace, we have

$$\begin{aligned} tr(M) &= tr[I_n - X(X'X)^{-1}X'] \\ &= n - tr[X(X'X)^{-1}X'] \\ &= n - tr[X'X(X'X)^{-1}] \\ &= n - tr(I_k) \\ &= n - k \end{aligned} \tag{4.22}$$

As a result, we obtain

$$\frac{e'e}{\sigma^2} \sim \chi_{n-k}^2 \tag{4.23}$$

Since $e'e = (n-k)S^2$, we finally obtain

$$\frac{(n-k)S^2}{\sigma^2} \sim \chi_{n-k}^2 \tag{4.24}$$

The result in Equation (4.24) allows us to establish a confidence interval for the unknown σ^2 as follows:

$$\Pr(\chi_{1-\alpha/2}^2 \leq \chi^2 \leq \chi_{\alpha/2}^2) = 1 - \alpha \tag{4.25}$$

where the χ^2 value in the middle of the double inequality is as shown in Equation (4.24) and where $\chi_{1-\alpha/2}^2$ and $\chi_{\alpha/2}^2$ are two critical chi-square values obtained from the chi-square table for $(n-k)df$ in such a manner that they

[4]Stewart, J., & Gill, L. (1991). *Econometrics* (2nd ed., p. 42). Harlow, England: Prentice Hall.

Figure 4.2   The Chi-Square Distribution

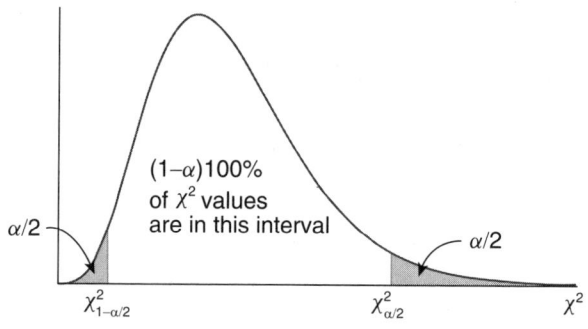

cut off a $100(\alpha/2)\%$ tail of the chi-square distribution (remember that the chi-square distribution is skewed; see Figure 4.2).

Substituting χ^2 from Equation (4.24), we obtain

$$\Pr\left[(n-k)\frac{S^2}{\chi^2_{\alpha/2}} \leq \sigma^2 \leq (n-k)\frac{S^2}{\chi^2_{1-\alpha/2}}\right] = 1-\alpha \qquad (4.26)$$

which gives the $100(1-\alpha)\%$ confidence interval for σ^2.

For the wage regression, we find from Table 4.1 that S^2 is about 42.3558 ($\approx$ square of 6.508137). We leave it to the reader to establish a 95% confidence interval for the true variance of the wage regression.

4.8.1 A Caution on the Confidence Interval Approach to Hypothesis Testing

According to Hendry and Nielsen,[5] "The trouble with confidence intervals is that, although they are useful, we cannot easily work with them using probability theory. A formal theory can instead be formulated for statistical tests . . .", as shown in Section 4.9.

4.9 Large-Sample Tests of Hypotheses

So far, we have discussed several specific hypothesis tests in the context of linear (in-parameter) regression models. Now, we discuss some broader

---

[5]Hendry, D. F., & Nielsen, B. (2007). *Econometric modeling: A likelihood approach* (p. 24). Princeton, NJ: Princeton University Press. See also Weakliem, D. L. (2016). *Hypothesis testing and model selection in the social sciences* (pp. 51–53, 72–73). New York, NY: Guilford Press.

approaches to hypothesis testing that can handle regression models, whether they are linear in their parameters or nor. These approaches are (1) the **likelihood ratio (LR) test**, (2) the **Wald test**, and (3) the **Lagrange multiplier (LM) test**. But all these tests are large-sample tests and they are all based on the likelihood principle.

These approaches also take into account some of the limitations of the Neyman–Pearson (N–P) approach to hypothesis testing. With N–P, we can construct the most powerful critical regions for testing a simple null hypothesis against a simple alternative hypothesis. But for composite hypotheses, there is usually no unambiguous "best" test, and that is part of the reason that there are multiple contenders, such as the LR, Wald, and LM tests. Besides, the three tests can handle multiple hypotheses. *The three approaches are asymptotically equivalent in that the statistic associated with each of these tests follows the chi-square distribution.*

4.9.1 The Likelihood Ratio Test

We introduced the concept of the (log) likelihood function earlier in the chapter. The best way to describe the LR test is to refer to the wage regressions given in Tables 4.1 and 4.3. We call the regression in Table 4.1 the unrestricted regression, or the *long regression*, and the one in Table 4.3 the restricted regression, or the *short regression*, for it omits the variables non-white and union workers. For the unrestricted wage regression, we find from Table 4.1 that the value of the log-likelihood function is −4240.370, and from Table 4.3, we find the value of the log-likelihood function for the restricted regression is −4246.864. We call these functions U*l* and R*l*, the unrestricted and restricted log-likelihood functions, respectively. Now we form the following statistic:

$$\lambda = 2(\mathrm{U}l - \mathrm{R}l) \sim \chi_r^2 \qquad (4.27)^6$$

where r is the number of restrictions.

If the sample size is large, technically infinite, it can be shown that under the null hypothesis the restrictions are valid; the statistic λ follows the chi-square distribution with *df* equal to the number of restrictions imposed by the null hypothesis, 2 in the present case.[7]

---

[6]This expression can also be written as $-2(\mathrm{R}l - \mathrm{U}l)$ or as $-2\ln(R/U)$.

[7]The derivation of the distribution of λ is rather complicated. The interested reader may refer to Stewart, J., & Gill, L. (1991). *Econometrics* (2nd ed., pp. 129–133). Harlow, England: Prentice Hall. These pages discuss the mathematics behind the LR, Wald, and LM tests.

Turning to our example, we obtain

$$\lambda = 2[-4240.370 - (-4246.864)] = 12.988 \approx 13$$

From the chi-square table, we find that for 2 df the probability that a chi-square value of as much as 13 or greater is less than 0.0015, or less than 0.15%. Since this is a small probability, we can conclude that we should not omit the two variables excluded from the restricted wage regression. More positively, these two variables belong in the wage regression.

One drawback of the LR test is that it involves estimating both the unrestricted and restricted regressions. But the Wald and LM tests involve estimating only one regression, although their approaches are different. Before we discuss these alternative tests, Table 4.4 shows the results of the LR test using the *Eviews* statistical package. The results are about the same as we obtained by hand. The output also presents the result of the restricted regression.

4.9.2 The Wald Test

This test is named after Abraham Wald (1902–1950), an influential mathematical statistician who died in a plane crash in India in 1950. Unlike the LR test, the Wald test estimates only one model, the unrestricted regression, but then tests the null hypothesis that one or more regression coefficients are zero or some linear restrictions of them. If this is the case, then there is no point in adding the "superfluous" variables to the model. The mathematics behind the Wald test is complicated. The basic idea behind the Wald test is that it tests how far from zero (or any other null hypothesis) the estimated parameters are in terms of their standard errors. *But unlike the standard regression output, the Wald test can be used to test multiple parameters simultaneously.*

In our restricted wage regression, the null hypothesis is that the coefficients of the regressors nonwhite and union are simultaneously equal to zero. Under the null hypothesis,

$$W \sim \chi_r^2 \tag{4.28}$$

where r is the number of restrictions, or the number of regressors excluded from the full model.

We use *Eviews 9* to conduct the Wald test, which gives the results shown in Table 4.5. Note that, like the LR test, the Wald statistic follows the chi-square distribution with the degrees of freedom equal to the number of restrictions imposed, 2 df in the present case.

In Table 4.5, $C(2)$ and $C(3)$ are the coefficients of nonwhite and nonunion. As the results show, the estimated χ^2 value is highly significant. Hence, we reject the restricted wage regression.

Table 4.4 Likelihood Ratio Test for the Wage Regression

Redundant Variables Test
Equation: UNTITLED
Specification: W C FE NW UN ED EX
Redundant Variables: NW UN

	Value	df	Probability
F-statistic	6.4957	(2, 1283)	0.0015599978933631008
Likelihood ratio	12.9865	2	0.0015135526954576009

F-test summary:

	Sum of Sq.	df	Mean Squares
Test SSR	550.2648	2	275.13241604288
Restricted SSR	54892.8086	1285	42.71813899575555
Unrestricted SSR	263373	1283	42.35584078436267

LR test summary:

	Value	df	
Restricted LogL	−4246.8635	1285	
Unrestricted LogL	−4240.3702	1283	

79

Restricted Test Equation:

Dependent Variable: W

Method: Least Squares

Date: 06/03/15   Time: 12:52

Sample: 1 1289

Included observations: 1289

Variable	Coefficient	Std. Error	t-Statistic	Prob.
C	-7.6536	1.0059	-7.6075224212404952	5.3765432069936535e-14
FE	-3.1864	0.3644	-8.7436515164874l	6.9356929788715 61e-18
ED	1.39371	0.06584	21.168131938 36954	1.4719481357368 05e-85
EX	0.1747	0.0158	11.00114597783916	5.7694973692490 28e-27
R-squared	0.3164			12.3658494573312
Adjusted R-squared	0.3148			7.8963502996 61898
S.E. of regression	6.5359			6.595599053189651
Sum squared resid	54892.8086			6.611616499281005
Log likelihood	-4246.863			6.6016113820774l1
F-statistic	198.3316			1.9026392356 36926

Table 4.5 Wald Test

Wald Test:

Equation: Untitled

Test Statistic	Value	df	Probability
F-statistic	41.6049	(2, 1283)	0.1113
Chi-square	83.2099	2	0

Null Hypothesis: C(2)=0,C(3)=0

Null Hypothesis Summary:

Normalized Restriction (= 0)	Value	Std. Err.
C(2)	−3.07487546	0.3646
C(3)	−1.5653	0.5091

Restrictions are linear in coefficients.

Notice that Tables 4.4 and 4.5 report the F statistic beside the chi-square statistic. This is not accidental, for there is a relationship between the two statistics, which is

$$F(q,n) \rightarrow \frac{1}{q} \chi_q^2 \qquad (4.29)$$

as $n \rightarrow \infty$, where q is the numerator df and n is the denominator df. In the present instance, $q=2$, $n=1,283$. The reader can verify that the F values in Tables 4.4 and 4.5 are half the corresponding chi-square values. Also, note that in the standard LRM, the F test described previously is an LR test and the t test is the Wald test.

4.9.3 The Lagrange Multiplier or the Score Test

In the Wald test, we estimate the unrestricted regression and then find out if it is worth retaining one or more variables in that regression. In the LM test, on the other hand, we estimate the restricted regression and then find out if it is worth adding one or more variables in the original unrestricted regression. The simplest way to conduct the LM test involves four steps: (1) we estimate the restricted model; (2) we obtain the residuals from the first step, calling them e_r; (3) we regress residuals e_r on all the regressors included in the unrestricted regression; and (4) we get the R^2 value obtained in Step 3 and multiply it by the number of observations in the sample, n, to obtain $LM = nR^2$.

If the null hypothesis is true (i.e., the restricted model is valid), it can be shown that

$$LM = nR^2 \sim \chi_g^2 \qquad (4.30)$$

where g is the number of restrictions imposed, that is, the number of regressors omitted from the restricted regression. In words, LM computed as shown in Equation (4.30) follows the chi-square distribution with g degrees of freedom. In the present example, g is 2.

To save space, we will not produce all the details, except to note that regressing the residuals from the restricted regression on all the regressors in the full model gave an R^2 of 0.010024. Multiplying this value by $n = 1,289$, we obtain

$$LM = nR^2 = 12.93209 \sim \chi_2^2$$

This chi-square value for 2 df is highly significant, suggesting that we can reject the null hypothesis that the restricted regression in the present instance is the right one. In other words, the variables nonwhite and union belong in the wage regression.

In the linear model, the LM test is related to the F test as follows:

$$LM = \frac{ngF}{n - k + gF} \qquad (4.31)$$

where n is the sample size, k is the number of regressors in the model, and g is the number of restrictions imposed. For very large n, it can be shown that

$$LM \approx gF \qquad (4.32)$$

It may be noted that in nonlinear regression models, the LM test is sometimes easier to implement than the LR or Wald tests.

4.9.4 Relationships Among the Three Tests

For the LRM, there is an interesting relationship among the three tests, which is

$$W \geq LR \geq LM \qquad (4.33)[8]$$

[8]See Maddala, G. S., & Lahiri, K. (2009). *Introduction to econometrics* (4th ed., pp. 120–122). Chichester, England: Wiley. For a survey of the three tests, see Engle, R. F. (1984). Wald, likelihood ratio, and Lagrange multiplier tests in econometrics. In Z. Griliches & M. D. Intriligator (Eds.), *Handbook of econometrics* (Vol. 2). Amsterdam, The Netherlands: North Holland.

This suggests that a hypothesis can be rejected by the Wald test, but not rejected by the LM test, although this does not happen regularly in large samples. However, this does not mean that one test is better than the others.

4.9.5 Caution About the Three Tests

Although useful in large samples and in dealing with nonlinear in-parameter regression models, their use in linear in-parameter regression models may not be necessary. As Davidson and MacKinnon note,[9]

For LRMs, with or without normal errors, there is of course no need to look at LM, W and LR at all since no information is gained from doing so over and above what is already contained in F.

4.10 Summary

In this chapter, our focus was on testing hypotheses regarding the values of the parameters of the LRM. We first discussed the types of hypothesis—simple or composite. To test these hypotheses, we developed the concepts of a **test statistic**, such as the Z test, the t test, the F test, and the χ^2 test. We explained the rationale behind these tests; the underlying probability theory is discussed in Appendix D.

On the basis of one or more of these tests, we decide whether we want to reject or not reject a particular hypothesis. In making this decision, we are likely to make either a Type I error or a Type II error. Type I error is when we reject a null hypothesis even though it may be true. Type II error is not rejecting a false hypothesis. For a given sample size, it is not possible to minimize both types of errors. The traditional approach, contained in the N–P lemma, has been to fix a Type I error at a sufficiently low level and not worry too much about a Type II error, the underlying philosophy being that somehow a Type I error is more serious than a Type II error.

To illustrate the various tests, we used a concrete example about the determination of hourly wages for a sample of 1,289 workers in the United States. The dependent variable in this example is the hourly wages in dollars, and the explanatory variables are years of education, years of work experience, gender of the worker, race of the worker, and whether a worker belongs to a labor union. The latter three variables are qualitative variables, taking values of 1 and 0, depending on whether or not a worker possesses a particular attribute.

[9]Davidson, R., & MacKinnon, J. (1993). *Estimation and inference in econometrics* (p. 456). New York, NY: Oxford University Press.

Based on the statistical results, we tested a variety of hypotheses:

1. Whether a particular coefficient is statistically significant or not, the null hypothesis being that the relevant population coefficient is zero. To test this hypothesis, we use the t test.

2. Whether collectively all the (partial) regression coefficients are zero. To test this hypothesis, we use the F test.

3. To assess the incremental contribution of one or more regressors, we use the F test.

To test linear restrictions on parameters, we use the F test, either by direct substitution or by the constrained least-squares method using the LM method.

Instead of using the test of significance approach to test hypotheses, we can use the method of confidence interval. A confidence interval provides a range of values that has certain probability of including the true values of the parameters.

In this chapter, we also discussed three large-sample tests of hypothesis. These are the LR test, the Wald test, and the LM test. Asymptotically, these tests are equivalent and they all follow the χ^2 distribution. However, for LRMs, there is no particular advantage to these tests over the traditional F test.

Appendix 4A shows the restricted least squares with matrix algebra.

Exercises

4.1 In the classical bivariate LRM,

$$Y_i = B_1 + B_2 X_i + u_i$$

Show that

a. $\operatorname{cov}(b_1, b_2) = -\bar{X} \left(\dfrac{\sigma^2}{\Sigma x_i^2} \right)$

and

b. $\operatorname{cov}(\bar{Y}, b_1) = 0$

where $x_i = (X_i - \bar{X})$.

4.2 In the bivariate regression model of Exercise 4.1, show that

   a. $E(\text{MESS}) = \sigma^2 + b_2 \Sigma x_1^2$, and

   b. $E(\text{MRSS}) = \sigma^2$

where MESS = mean or average explained sum of squares and MRSS = mean or average residual sum of squares. (*Hint:* Refer to Table 4.2.)

4.3 Consider the data in Tables 1 and 2 below:

Table 1			Table 2		
Y	X_2	X_3	Y	X_2	X_3
1	2	4	1	2	4
2	0	2	2	0	2
3	4	12	3	4	0
4	6	0	4	6	12
5	8	16	5	8	16

The only difference between the two tables is that the third and fourth values of X_3 are interchanged.

   a. Estimate the two regressions.

   b. What conclusions do you draw from this exercise?

   c. What may be the reason for the difference between the two regressions?

4.4 Suppose in the model

$$Y_i = B_1 + B_2 X_{2i} + B_3 X_{3i} + u_i$$

r_{23}, the coefficient of correlation between X_2 and X_3 is zero. Therefore, someone suggests that you estimate the following regressions:

$$Y_i = C_1 + C_2 X_{2i} + u_{1i}$$
$$Y_i = D_1 + D_3 X_{3i} + u_{2i}$$

   a. Will $c_2 = b_2$ and $d_3 = b_3$? Why?

   b. Will b_1 equal to c_1 or d_1, or some combination thereof?

   c. Will $\text{var}(b_2) = \text{var}(c_2)$ and $\text{var}(b_3) = \text{var}(d_3)$? Explain.

4.5 Suppose that $X_1, X_2, \ldots, X_n$ is a random sample from $N(\gamma, 1)$. The ML estimator of γ is $\hat{\gamma} = \bar{X}$. What is the ML estimator of $T = e^{\gamma}$? (*Hint:* Consistency property of ML estimators.)

4.6 Suppose that $X_1, X_2, \ldots, X_n$ is a random sample from $N(\theta, \sigma^2)$. The ML estimator of θ is $\hat{\theta} = \bar{X}$. Let $\tilde{\theta}$, the sample median, be an alternative estimator of θ. It can be shown that the ML estimator satisfies[10]

$$\sqrt{n}(\hat{\theta}_n - \theta) \sim N(0, \sigma^2) \text{ (i.e., asymptotic normality)}$$

It can also be shown that the median satisfies

$$\sqrt{n}(\tilde{\theta} - \theta) \sim N\left(0, \ \sigma^2 \frac{\pi}{2}\right)$$

That is, it is also asymptotically normally distributed with the stated parameters.

Since both the estimators converge to the right value, what is the difference between the two? Which is relatively more efficient? What is the practical importance of this finding?

Appendix 4A: Constrained Least Squares: OLS Estimation Under Linear Restrictions[11]

Suppose we want to estimate the LRM

$$y = XB + u \tag{4A.1}$$

subject to the linear restriction

$$RB = r \tag{4A.2}$$

where R is a $q \times k$ matrix of *known constants*, with $q < k$ and r is a q vector of *known constants*. The null hypothesis in the present instance is

$$H_0 : RB - r = 0 \tag{4A.3}$$

---

[10]See Wasserman, L. (2005). *All of statistics: A concise course in statistical inference* (pp. 130–131). New York, NY: Springer.

[11]See Johnston, J., & Dinardo, J. (1997). *Econometric methods* (4th ed., pp. 103–104). New York, NY: McGraw-Hill.

To shed light on this type of construct, refer to the wage regressions given in Tables 4.1 and 4.2. In the latter, we dropped the variables nonwhite and union. So we can write the constraint in (4A.3) as

$$R = \begin{bmatrix} 0 & 0 & 1 & 0 & 0 \\ 0 & 0 & 0 & 1 & 0 \end{bmatrix} \quad r = \begin{bmatrix} 0 \\ 0 \end{bmatrix} \tag{4A.4}$$

Here R is 2×5 and r is 2×1.

Written out fully, Equation (4A.2) now becomes

$$R = \begin{bmatrix} 0 & 0 & 1 & 0 & 0 \\ 0 & 0 & 0 & 1 & 0 \end{bmatrix} \begin{bmatrix} B_1 \\ B_2 \\ B_3 \\ B_4 \\ B_5 \end{bmatrix} \quad r = \begin{bmatrix} 0 \\ 0 \end{bmatrix} \tag{4A.5}$$

To estimate (4A.1) subject to the linear equality restriction (4A.2), we form the following **Lagrangian function**, letting b^* be the OLS estimator of the restricted regression:

$$Z = (y - Xb^*)'(y - Xb^*) - 2\lambda'(Rb^* - r) \tag{4A.6}$$

where λ is a q vector of LMs. To minimize (4A.6), the first-order conditions are as follows:

$$\frac{\partial Z}{\partial b^*} = -2X'y + 2X'Xb^* - 2R'\lambda = 0 \tag{4A.7}$$

$$\frac{\partial Z}{\partial \lambda} = -2(Rb^* - r) = 0 \tag{4A.8}$$

Solving, we obtain

$$(X'X)b^* = X'y + R'\lambda \tag{4A.9}$$

$$Rb^* = r \tag{4A.10}$$

Equation (4A.9) yields

$$b^* = (X'X)^{-1}X'y + (X'X)^{-1}R'\lambda$$
$$= b + (X'X)^{-1}R'\lambda \tag{4A.11}$$

where b is the (usual) unrestricted estimator of B.

The constrained estimator is thus equal to the unconstrained estimator plus a linear combination of the multipliers. The two estimators will coincide if and only if λ is zero, in which case, the constraint is not binding, that is, it is satisfied in the data without imposing the construct. Premultiplying (4A.11) by R gives

$$Rb^* = Rb + [R(X'X)^{-1}R']\lambda \qquad (4A.12)$$

which can be written as

$$Rb^* - Rb = [R(X'X)^{-1}R']\lambda \qquad (4A.13)$$

Solving for λ, we obtain

$$\lambda = [R(X'X)^{-1}R']^{-1}(Rb^* - Rb)$$
$$= [R(X'X)^{-1}R']^{-1}(r - Rb) \qquad (4A.14)$$

Substituting this in (4A.11), we obtain

$$b^* = b + (X'X)^{-1}R'[R(X'X)^{-1}R']^{-1}(r - Rb) \qquad (4A.15)$$

It is evident that the difference between the restricted and the unrestricted LS coefficient vector is a linear function of the vector $r - Rb$, which is an indication of the degree to which the unconstrained estimator fails to satisfy the constraint. Now if the (unconstrained) OLS estimator b satisfies the constraint $r = Rb$, then $b^* = b$, even if we did not impose the restriction. If the restriction is not satisfied, b^* will be biased.

The residuals from the restricted regression are given by[12]

$$e^* = y - Xb^*$$
$$= (y - Xb) - X(b^* - b)$$
$$= e - X(b^* - b) \qquad (4A.16)$$

Transposing (4A.16) and multiplying, we obtain

$$e^{*\prime}e^* = e'e + (b^* - b)' X'X(b^* - b) \qquad (4A.17)$$

---

[12]For further details, see Davidson, J. (2000). *Econometric theory* (pp. 28–30). Malden, MA: Blackwell. See also Darnell, A. C. (1994). *A dictionary of econometrics* (pp. 348–351). Cheltenham, England: Edward Elgar.

Substituting for $(b^* - b)$ from (4A.15) and simplifying, we obtain

$$e^{*'}e^* - e'e = (r - Rb)'[R(X'X)^{-1}R']^{-1}(r - Rb) \qquad (4A.18)$$

To test the hypothesis that $Rb = r$, we can now use the F statistics as follows:

$$F = \frac{(e^{*'}e^* - e'e)/q}{e'e/(n-k)} \sim F(q, n-k) \qquad (4A.19)$$

As usual, if the computed F is not statistically significant, we will not reject the null hypothesis, but if it is significant, we will reject the hypothesis, in which case the imposed restriction(s) is not valid.

CHAPTER 5. GENERALIZED LEAST SQUARES (GLS): EXTENSIONS OF THE CLASSICAL LINEAR REGRESSION MODEL

5.1 Introduction

Given the assumptions of the CLRM, we have shown that the OLS estimators are BLUE, that is, they are efficient. With the added assumption of normality for the error term, the OLS estimators are efficient in the entire class of unbiased estimators, whether linear or nonlinear. Efficiency is a measure of the variance of an estimator's sampling distribution—the smaller the variance, the better the estimator. What happens if one or more assumptions of the CLRM are not fulfilled?

In this chapter, therefore, we go beyond the "comfortable" world of the CLRM and consider a more general framework that allows us to relax one or more assumptions of the CLRM, assumptions that may not be tenable in practice.

We start with a more general setup:

$$y = XB + u \tag{5.1}$$

where y is an $n \times 1$ vector of observations on the dependent variable, or the regressand; X is an $n \times k$ matrix of n observations on k regressors, one of which is an intercept taking the value of 1 for all observations; B is a $k \times 1$ vector of coefficients of the k regressors; and u is an $n \times 1$ vector of n error terms.

We assume that $E(u \mid X) = 0$, as before, and also assume that the X regressors are fixed numbers (in Chapter 6, we will relax this assumption). But instead of assuming that

$$E(uu') = \sigma^2 I \tag{5.2}$$

as in the classical linear model, we now assume that

$$E(uu') = \sigma^2 V \tag{5.3}$$

where V is a PD matrix and its elements are assumed to be known (we will relax this assumption shortly).[1]

---

[1] A symmetric matrix A is positive-definite if $x'Ax > 0$ for all nonzero x. A is a positive semidefinite matrix if $x'Ax \geq 0$ for all x, and there is at least one nonzero x for which $x'Ax = 0$. (For further details, see Appendix A).

Equation (5.2) is called a **scalar covariance matrix**,[2] for it is the product of a scalar and an identity matrix. This structure of covariance matrix implies that the regression error variance is homoscedastic and that there is no pairwise correlation among the error terms (i.e., there is no autocorrelation). Such a scalar covariance matrix may not be appropriate in all situations. This is why we consider the more general variance–covariance matrix given in Equation (5.3), which is called a **nonscalar covariance matrix**.

Written out fully, Equation (5.3) becomes

$$E(\boldsymbol{uu'})=\sigma^2 V$$

$$=\sigma^2 \begin{bmatrix} E(u_1^2) & E(u_1 u_2) & E(u_1 u_3) & E(u_1 u_4) & \cdots & E(u_1 u_n) \\ E(u_2 u_1) & E(u_2^2) & E(u_2 u_3) & E(u_2 u_4) & \cdots & E(u_2 u_n) \\ E(u_3 u_1) & E(u_3 u_2) & E(u_3^2) & E(u_3 u_4) & \cdots & E(u_3 u_n) \\ \vdots & & & \ddots & \\ \vdots & & & & \ddots \\ E(u_n u_1) & E(u_n u_2) & E(u_n u_3) & E(u_n u_4) & \cdots & E(u_n^2) \end{bmatrix} \quad (5.4)$$

Note: $E(u_i \mid X)=0$ for each observation.

This can also be expressed as

$$\mathrm{cov}(\boldsymbol{u})=E(\boldsymbol{uu'})$$

$$= \begin{bmatrix} \sigma_1^2 & \sigma_{12} & \sigma_{13} & \sigma_{14} & \cdots & \sigma_{1n} \\ \sigma_{21} & \sigma_2^2 & \sigma_{23} & \sigma_{24} & \cdots & \sigma_{2n} \\ \sigma_{31} & \sigma_{32} & \sigma_3^2 & \sigma_{34} & \cdots & \sigma_{3n} \\ \vdots & & & \ddots & & \vdots \\ \vdots & & & & \ddots & \vdots \\ \sigma_{n1} & \sigma_{n2} & \sigma_{n3} & \sigma_{n4} & \cdots & \sigma_n^2 \end{bmatrix} \quad (5.5)$$

where the entries on the main diagonal are variances of the n error terms and $\sigma_{ij}\,(i \neq j)$ are the pairwise covariances among the error terms. Notice that like Equation (5.4), Equation (5.5) is also symmetrical.

In this setup, the entries along the main diagonal give the variance of each error term, thus allowing for **heteroscedastic**, or unequal, variances, and the entries off the main diagonal give the pairwise covariances among

---

[2]It is often called a variance–covariance matrix, but we use the latter term, for variance is simply the covariance of a variable with itself.

the error terms, thus allowing for the possibility of **autocorrelation**. If all the off-diagonal elements of V are zero, Equation (5.5) will represent no autocorrelation but heteroscedastic variances. If the off-diagonal elements are nonzero, there is autocorrelation. If, in addition, all the entries on the main diagonal are 1, and all other entries are zero, we are back to the scalar covariance matrix in Equation (5.2), in which case we are back to the classical linear regression setup.

Figure 5.1 shows both homoscedasticity and heteroscedasticity for a bivariate regression of savings on income. Figure 5.1a shows that as income increases, the average level of savings increases, but the variability of savings as measured by σ^2 remains the same. This is the case of homoscedasticity. Figure 5.1b, on the other hand, shows that as income increases, the average level of savings also increases and so does the variability of savings. This is the case of heteroscedasticity.

Figure 5.1   (a) Homoscedastic Disturbances and (b) Heteroscedastic Disturbances

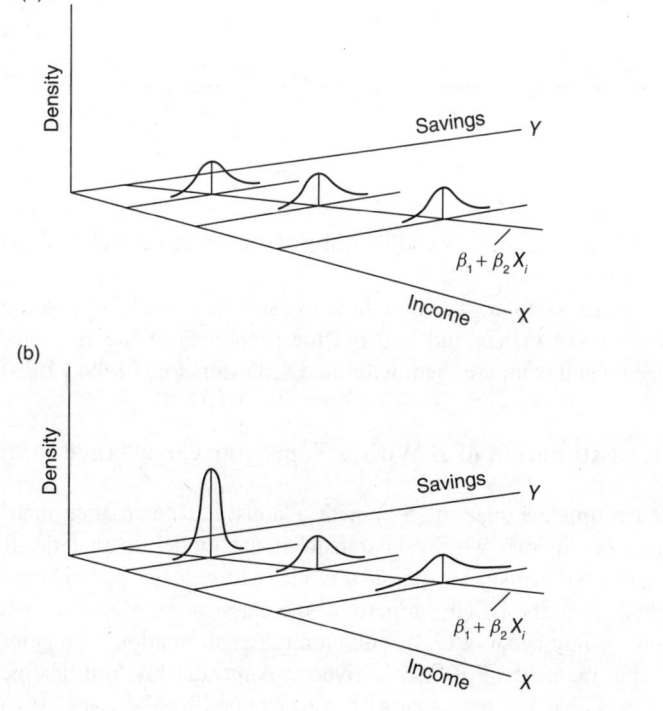

Figure 5.2   (a) Positive Autocorrelation and (b) Negative Autocorrelation

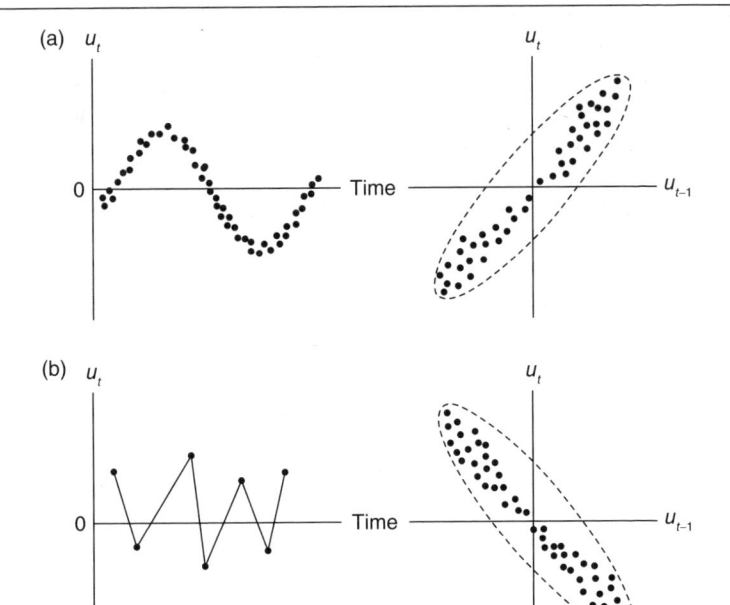

Figure 5.2 shows hypothetical patterns of autocorrelation in the regression error term, u_t.

In this chapter, we show how to estimate model (5.1) with a nonscalar covariance matrix and find out the properties of the estimated **B** coefficients and compare them with the OLS estimators of the CLRM.

5.2 Estimation of B With a Nonscalar Covariance Matrix

To estimate Equation (5.1) with a nonscalar covariance matrix, we can proceed in two ways: (1) transform the model so that the transformed model will satisfy the classical model with a scalar covariance matrix, and thus retain the BLUE property of the classical model, or (2) estimate (5.1) taking into account (5.3) without any transformation. The latter approach, using the method of ML, is given in Appendix 5A. But this method gives results that are the same as those obtained by the transformation of the model, especially in large samples.

5.2.1 Estimation of LRM by Transformation

First, note that we are now assuming

$$E(u \mid X) = 0$$
$$\text{cov}(u) = E(uu') = \sigma^2 V \qquad (5.6)^3$$

Since V is a symmetric PD matrix, its inverse V^{-1} is also PD, which means we can apply Choleski decomposition to V^{-1} and find a nonsingular lower triangular matrix P such that

$$V^{-1} = P'P \qquad (5.7)^4$$

where

$$V = (P'P)^{-1}$$
$$= P^{-1}(P')^{-1} \qquad (5.8)$$

Therefore,

$$PVP' = I \qquad (5.9)$$

Now, premultiplying (5.1) by P, we obtain

$$Py = PXB + Pu \qquad (5.10)$$

The transformed error term, Pu, has the following properties:

$$E(Pu) = PE(u) = 0 \qquad (5.11)$$
$$\text{var}(Pu) = P\text{var}(u)P'$$
$$= P(\sigma^2 V)P'$$
$$= \sigma^2 PVP'$$
$$= \sigma^2 I, \quad \text{using (5.9)} \qquad (5.12)$$

The transformed model (5.10) has a scalar covariance matrix—no heteroscedasticity and no autocorrelation. As a result, estimation of (5.10)

---

[3]Some textbooks express the nonscalar covariance matrix as $E(uu') = \Omega$, without the scalar σ^2. But we will keep it just in case we transform the model that makes the cov(u) a scalar covariance matrix, as in the classical model. Or we can treat it as a proportionality factor in the case of heteroscedasticity.

[4]If a square matrix V is nonsingular, then we can find a nonsingular triangular matrix Z such that $ZZ' = V$. The matrix Z is often called the **square root matrix** for V. This procedure is called Choleski decomposition.

by OLS will produce estimators that are BLUE. These estimators and their variances are as follows (see the similarity with the usual OLS estimators):

$$
\begin{aligned}
b^{gls} &= \left[(PX)'(PX)\right]^{-1}(PX)'Py \\
&= (X'P'PX)^{-1}X'P'Py \\
&= (X'V^{-1}X)X'V^{-1}y
\end{aligned}
\tag{5.13}
$$

$$
\begin{aligned}
\operatorname{cov}(b^{gls}) &= \sigma^2\left[(PX)'(PX)\right]^{-1} \\
&= \sigma^2(X'V^{-1}X)^{-1}
\end{aligned}
\tag{5.14}
$$

where b^{gls} is the estimator of the regression parameters obtained by the method of generalized least squares (GLS). It is easy to show that b^{gls} is an unbiased estimator of B. (*Hint:* Put $y = XB + u$ into (5.13) and take the expectation of b^{gls}.)

Note that both these equations reduce to the standard OLS equations if $V = I$. The easiest way to remember Equations (5.13) and (5.14) is to interpose V^{-1} in the usual OLS equations in an obvious way. Also note that b^{gls} is a linear estimator, as it is a linear function of y.

If we assume the error term to be normally distributed with zero mean and constant variance, then we know that

$$
b \sim N(B, \sigma^2 I)
\tag{5.15}
$$

Since GLS is simply OLS applied to the transformed model (5.10), and if we assume that u is normally distributed, then

$$
b^{gls} \sim N[B, \sigma^2(X'V^{-1}X)^{-1}]
\tag{5.16}
$$

Therefore, *under the normality assumption, b^{gls} is also the ML estimator. It is also BLUE.* However, note that b^{gls} is BLUE whether normality is assumed or not, but with the normality assumption it is a **uniformly minimum variance unbiased estimator (UMVUE)**.

Recall that for the standard OLS, the error variance is estimated as

$$
S^2 = \frac{(y - Xb)'(y - Xb)}{n - k} = \frac{e'e}{n - k}
\tag{2.18}
$$

For the GLS, the equivalent formula can be computed from the GLS residuals as

$$
e_{gls} = y - Xb^{gls}
\tag{5.17}
$$

and the estimated GLS error variance is as follows:

$$S_{\text{gls}}^2 = \frac{e'_{\text{gls}} V^{-1} e_{\text{gls}}}{n-k} \qquad (5.18)^5$$

Given this information, we can conduct hypothesis testing and develop confidence intervals in the usual (OLS) manner. No new principles are involved.

Note that if V is a scalar covariance matrix, then GLS collapses to OLS. It is therefore easy to understand why b^{gls} is called GLS. It may be noted that b^{gls} is also known as the **generalized Aitken estimator**, following Aitken's generalization of the Gauss–Markov theorem to the general LRM.[6]

5.2.2 OLS Versus GLS

In the case of OLS, we have shown that b, the estimator of B, is unbiased. In the case of GLS also, the estimator b^{gls} is unbiased, which can be seen easily from Equation (5.13) by taking expectations of this equation on both sides and noting that V is known and that X is nonstochastic. In the case of a nonscalar covariance matrix, it is GLS that is BLUE and not the OLS. The difference between the two lies in their respective covariance matrices. The GLS covariance matrix is given in Equation (5.14).

If you examine the derivation of cov(b) under the classical assumptions given in Equation (2.8), you will see that the fourth line in that equation has the term $E(uu')$, but this expectation now is the matrix V and not $\sigma^2 I$. As a result, we obtain

$$\text{cov}(b) = \sigma^2 (X'X)^{-1} X'VX(X'X)^{-1} \qquad (5.19)$$

whereas cov(b^{gls}) is as given in Equation (5.14).

On the other hand, cov(b) $= \sigma^2(X'X)^{-1}$, which is obviously different from (5.19), unless $V = I$.

The difference between cov(b) given in (5.19) and the GLS covariance given in (5.14) is

$$\begin{aligned}
\text{cov}(b) - \text{cov}(b^{\text{gls}}) &= \sigma^2 (X'X)^{-1} X'VX(X'X)^{-1} - \sigma^2 (X'V^{-1}X)^{-1} \\
&= \sigma^2 AVA'
\end{aligned} \qquad (5.20)$$

---

[5]For details, see Ravishankar, N., & Dey, D. K. (2002). *A first course in linear model theory* (pp. 122–124). New York, NY: Chapman & Hall; Theil, H. (1971). *Principles of econometrics* (pp. 238–240). New York, NY: Wiley.

[6]Aitken, A. C. (1935). On least squares and linear combination of observations. *Proceedings of the Royal Society of Edinburgh, 55,* 42–48.

where

$$A = (X'X)^{-1}X' - (X'V^{-1}X)^{-1}X'V^{-1} \tag{5.21}$$

which is a PD matrix. This means that the OLS cov(b), even taking into account the covariance matrix V, is larger than the covariance matrix of b^{gls}. Only in special cases will the two covariance matrices be equal.[7] The importance of this finding is that if we use the scalar covariance matrix routinely in situations where it may be inappropriate, we will be estimating standard errors of the regression coefficients incorrectly and, therefore, inference based on the application of the t and F tests will be misleading.

One big question still remains: We have assumed that V is known. But this is rare. What then?

5.3 Estimated Generalized Least Squares[8]

Since V is generally not known, one suggested alternative is to estimate it. One simple way to estimate it is to run OLS and obtain the residuals, $e = [e_1, e_2, \ldots, e_n]'$.

From these residuals, we obtain the covariance matrix, cov(e), call it $\hat{V}$, and use it in lieu of V in Equation (5.14). Note that the expected value of each residual is zero (see Equation (2.12)).

The estimated B now is

$$b^{egls} = (X'\hat{V}^{-1}X)^{-1}X'\hat{V}^{-1}y \tag{5.22}$$

And its covariance matrix is

$$\text{cov}(b^{egls}) = (X'\hat{V}^{-1}X)^{-1} \tag{5.23}$$

where "egls" denotes estimated generalized least square (EGLS).

5.3.1 Properties of EGLS

If V is known, the GLS estimator b^{gls} is BLUE. But if we use $\hat{V}$, the EGLS coefficients are not BLUE in finite samples. However, if the sample size is large (technically infinite), $\hat{V}$ is a consistent estimator of V, in which

---

[7]See Amemiya, T. (1994). *Introduction to statistics and econometrics* (chap. 13). Cambridge, MA: Harvard University Press. Usually, the strict inequality holds. See also Judge, G. G., Carter Hill, R., Griffiths, W. E., Lutkepohl, H., & Lee, T.-C. (1988). *Introduction to the theory and practice of econometrics* (2nd ed., chap. 8). New York, NY: Wiley.

[8]It is also known as **feasible generalized least squares (FGLS)**.

case the difference between b^{gls} and b^{egls} will not be significant. Although the estimated V provides asymptotically correct standard errors, they are still inefficient relative to the (true) GLS.

In practice, it may be possible to assume some specialized form of V, pertaining to the nature of heteroscedasticity and autocorrelation, which will make GLS estimation easier. We now consider some specialized V matrices.

5.4 Heteroscedasticity and Weighted Least Squares

If there is heteroscedasticity but no autocorrelation, $\text{var}(u_i) = \sigma_i^2$, the variance–covariance matrix V assumes the following form:

$$\text{cov}(V) = E(uu')$$

$$= \begin{pmatrix} \sigma_1^2 & 0 & 0 & \dots & 0 \\ 0 & \sigma_2^2 & 0 & \cdots & 0 \\ 0 & 0 & \sigma_3^2 & \dots & 0 \\ \vdots & & & \ddots & \vdots \\ 0 & 0 & 0 & \dots & \sigma_n^2 \end{pmatrix} \qquad (5.24)$$

This is a nonscalar covariance matrix. In this setup, it is assumed that the error terms are uncorrelated, but they do not have a common variance. That is, the variances are heteroscedastic. This situation occurs frequently in regression analysis with cross-sectional data where the regressand depends on the magnitude of the regressors, such as consumption in relation to income in a cross-sectional study of family consumption–income analysis. In the presence of heteroscedasticity, the OLS estimators are unbiased, but they are not efficient. Therefore, the standard statistical hypothesis tests are not reliable, if we do not correct for heteroscedasticity.

If we knew σ_i^2 (a big if), we could obtain BLUE estimators by dividing each observation by the (heteroscedastic) σ_i and estimate the transformed model by OLS. This method of estimation is known as the method of **weighted least squares (WLS)**. In WLS, an observation drawn from a population with smaller variance is given more weightage than an observation drawn from a population with more variability. WLS is thus a special form of GLS. In WLS, we minimize the RSS, which takes the following form:

$$\text{RSS} = (y - Xb^w)'V^{-1}(y - Xb^w)$$

$$= \sum_i^n \left(\frac{Y_i - \hat{Y}_i}{\sigma_i} \right)^2 \qquad (5.25)$$

Note: "w" in b^w stands for weighted b, a form of GLS.

As this equation shows, each residual is weighted by $1/\sigma_i$, which is the reciprocal of the standard deviation of the population from which the observation is drawn. The smaller the σ_i, the greater the weight that an observation receives. On the other hand, the larger the σ_i, the lesser the weight that an observation receives. Stated differently, the weight an observation in the sample receives varies inversely with the standard deviation from the population to which that observation belongs.

Although elegant, the practical problem in the use of WLS is that we generally do not know σ_i. In practice, one makes some ad hoc assumption about σ_i^2 and transforms the data accordingly. For example, one can assume that

$$\sigma_i^2 \propto \sigma^2 X_i^2 \tag{5.26}$$

That is, the heteroscedastic variance varies in proportion to the square of one of the regressors in the model. If this assumption is correct, one can then transform the original model by dividing it through by X_i and use OLS to estimate the transformed model. Of course, there are several other transformations that one could use to achieve homoscedasticity.[9] There are also several diagnostic tests to detect heteroscedasticity, such as the Park test, the Glejser test, the Goldfeld–Quandt test, the Breusch–Pagan–Godfrey test, White's general heteroscedasticity test, and several graphic tests. Each of these tests has some limitations.[10]

As an example, suppose we have the following regression, call it the savings function:

$$Y_i = B_1 + B_2 X_i + u_i$$

where Y=family savings and X=family income. That is, the ith family's savings is a function of the ith family's income.

Suppose the variance of u_i is proportional to the square of family income, that is,

$$E(u_i^2) = \sigma^2 X_i^2$$

[9]For several such transformation, see Gujarati, D. N., & Porter, D. C. (2009). *Basic econometrics* (5th ed., chap. 11). New York, NY: McGraw-Hill.

[10]All these tests as well as their limitations are discussed in Gujarati, D. N., & Porter, D. C. (2009). *Basic econometrics* (5th ed., chap. 11). New York, NY: McGraw-Hill. See also Kaufman, R. L. (2013). *Heteroskedasticity regression: Detection and correction.* Thousand Oaks, CA: Sage. There is a debate in the literature as to whether it is homoscedasticity or homoskedasticity and heteroscedasticity or heteroskedasticity. Both seem to be acceptable.

Now if we divide the savings function by income on both sides of that function, we obtain

$$\frac{Y_i}{X_i} = B_1 \frac{1}{X_i} + B_2 + \frac{u_i}{X_i}$$

$$= B_1 \frac{1}{X_i} + B_2 + v_i$$

where $v_i = u_i / X_i$

Now,

$$E(v_i^2) = E(u_i / X_i)^2 = \frac{1}{X_i^2} E(u_i^2) = \frac{1}{X_i^2} (\sigma^2 X_i^2) = \sigma^2$$

As you can see, it is the transformed error term v_i that is homoscedastic.

This is an example of WLS; here, we weight each observation by the inverse of the income level. Of course, there are several other transformations that can be used. The details of the various transformations can be found in the references cited earlier.

5.5 White's Heteroscedasticity-Consistent Standard Errors[11]

We showed in (5.19) the covariance matrix of the OLS estimator using the V matrix, which we can write more explicitly as follows (suppressing the σ^2 factor for obvious reasons):

$$\text{cov}(b) = (X'X)^{-1} X'VX(X'X)^{-1}$$

$$\text{cov}(b) = (X'X)^{-1} X' \begin{pmatrix} \sigma_1^2 & 0 & 0 & \cdots & 0 \\ 0 & \sigma_2^2 & 0 & \cdots & 0 \\ 0 & 0 & \sigma_3^2 & \cdots & 0 \\ \vdots & & & & \\ 0 & 0 & 0 & \cdots & \sigma_n^2 \end{pmatrix} X(X'X)^{-1}$$

$$= (X'X)^{-1} \left(\sum_1^n \sigma_i^2 x_i x_i' \right) (X'X)^{-1} \tag{5.27}$$

where x_i is the ith row of the X matrix.

[11] White, H. (1980). A heteroscedasticity consistent covariance matrix estimator and a direct test of heteroscedasticity. *Econometrica, 48,* 817–838.

Since $E(u_i^2) = \sigma_i^2$, it is reasonable to estimate σ_i^2 by u_i^2. However, u_i is not observable. White suggests estimating u_i^2 by its sample counterpart, e_i^2, the squared residuals. The White procedure thus involves *two steps*. In Step 1, we estimate the LRM by OLS and obtain the residuals e_i and square them up and estimate the covariance matrix V as follows:

$$\hat{V} = \begin{pmatrix} e_1^2 & \cdots & 0 \\ \vdots & \ddots & \vdots \\ 0 & \cdots & e_n^2 \end{pmatrix} \tag{5.28}$$

where $\hat{V}$ = estimated V.

Therefore, the estimated covariance matrix assumes the following form:

$$\text{cov}(b) = (X'X)^{-1} \hat{V} X (X'X)^{-1}$$

Or more explicitly,

$$\text{cov}(b) = (X'X)^{-1} X' \begin{pmatrix} e_1^2 & \cdots & 0 \\ \vdots & \ddots & \vdots \\ 0 & \cdots & e_n^2 \end{pmatrix} X (X'X)^{-1} \tag{5.29}$$

which can also be expressed as

$$\text{estVar}(b) = (X'X)^{-1} \left(\sum_{i=1}^{n} e_i^2 x_i x_i' \right) (X'X)^{-1} \tag{5.29a}$$

where "est" means estimated.

Several statistical packages now include White's heteroscedasticity-corrected standard errors.

For the wage model presented in Table 4.1, heteroscedasticity-corrected standard errors are presented in Table 5.1. Such standard errors are also known as **robust standard errors**. They are called *robust* because they are asymptotically (i.e., in large samples) valid in the presence of any form of heteroscedasticity as well as homoscedasticity. *In general, robust estimation concerns the construction of alternative estimators that are less sensitive to violations of the assumptions of the CLRM.* However, a robust estimator is not the best if the assumptions of the classical linear regression are all met.

The first thing to notice about this output is that the coefficients of the variables given in Tables 4.1 and 5.1 are identical, but the standard errors

Table 5.1 Heteroscedasticity-Corrected Robust Standard Errors

Dependent Variable: W
Method: Least Squares
Date: 05/13/13 Time: 22:02
Sample: 1 1289
Included observations: 1289
White heteroskedasticity-consistent standard errors & covariance

Variable	Coefficient	Std. Error	t-Statistic	Prob.
C	−7.183338	1.090064	−6.589834	0.0000
FE	−3.074875	0.364256	−8.441521	0.0000
NW	−1.565313	0.397626	−3.936647	0.0001
UN	1.095976	0.425802	2.573908	0.0102
ED	1.370301	0.083485	16.41372	0.0000
EX	0.166607	0.016049	10.38134	0.0000

R-squared	0.323339	Mean dependent var	12.36585
Adjusted R-squared	0.320702	S.D. dependent var	7.896350
S.E. of regression	6.508137	Akaike info criterion	6.588627
Sum squared resid	54342.54	Schwarz criterion	6.612653
Log likelihood	−4240.370	Hannan-Quinn criter.	6.597646
F-statistic	122.6149	Durbin-Watson stat	1.897513
Prob(F-statistic)	0.000000	Wald F-statistic	100.8747
Prob(Wald F-statistic)	0.000000		

and the t statistics are different. Thus, *the White standard error correction only corrects the standard errors.* As Table 5.1 shows, some standard errors have decreased and some have increased, but there is not a very substantive change in the results in the two tables. It may be noted that the t statistics reported in Table 5.1 are known as **heteroscedasticity-robust t statistics**.

5.6 Autocorrelation

A commonly encountered problem in regression analysis involving time-series data is autocorrelation, correlation between or among the (time-series) observations, which is often reflected in the regression error term. Some patterns of autocorrelation are shown in Figure 5.2. The CLRM assumes that such correlations are absent.

The consequences of autocorrelation are similar to those of heteroscedasticity. Specifically, in the presence of autocorrelation, the OLS estimators are still unbiased and consistent; they are still normally distributed in large samples, but they are no longer BLUE. In most cases, OLS standard errors are underestimated, which means the estimated t values are inflated, giving the appearance that one or more regression coefficients are more significant than they actually might be. As a consequence, the usual t and F tests may not be valid. All these consequences as well as remedial measures to solve the autocorrelation problem are discussed in most introductory textbooks.[12]

The solution to autocorrelation depends on the nature of autocorrelation. Suppose in the LRM, $y = BX + u$, the error term u assumes the following form:

$$u_t = \rho_1 u_{t-1} + \rho_2 u_{t-2} + \cdots + \rho_p u_{t-p} + \varepsilon_t \tag{5.30}$$

Model (5.30), called an **autoregressive process of the pth order [AR(p)]**, states that the error term in time period t depends on the error terms in periods going back to p time periods: $u_{t-i}, i = 1, 2, \ldots, p$. The error term ε_t is a random error term that satisfies the usual OLS assumptions. The u_{t-i} are called the **lagged error terms**, the value of the maximum lag depending on the nature of the data: Time-series observations can be observed at various time periods, such as a day, a month, a year, or even much shorter time periods for data involving financial time series, such as exchange rates and interest rates. The name *autoregression* is appropriate, for it involves regression of the current value of a variable on its own past values.

The simplest of the AR(p) models is AR(1), which is

$$u_t = \rho u_{t-1} + \varepsilon_t \mid \rho \mid < 1 \tag{5.31}$$

Assuming that $E(\varepsilon_i) = 0, \mathrm{var}(\varepsilon_i) = \sigma_\varepsilon^2$, and $\mathrm{cov}(\varepsilon_i \varepsilon_j) = 0, i \neq j$, which are the standard assumptions of the CLRM, it can be shown that

$$\mathrm{var}(u_t) = \frac{\sigma_\varepsilon^2}{1 - \rho^2} \tag{5.32}$$

---

[12]See, for instance, Gujarati, D. (2015). *Econometrics by example* (2nd ed., chap. 6). Basingstoke, England: Palgrave-Macmillan; Gujarati, D. N., & Porter, D. C. (2009). *Basic econometrics* (5th ed., chap. 12). New York, NY: McGraw-Hill.

and

$$\text{cov}(\boldsymbol{u}) = \frac{\sigma_u^2}{1-\rho^2} \begin{pmatrix} 1 & \rho & \cdots & \rho^{t-1} \\ \vdots & & & \vdots \\ \rho^{t-1} & \rho^{t-2} & \cdots & 1 \end{pmatrix} = \frac{\sigma_u^2}{1-\rho^2} V \qquad (5.33)$$

where V is the matrix in the bracket.

To estimate the LRM under AR(1), we can use GLS as in (5.13), using the V, given in (5.33). But we again face the problem that ρ is unknown. We can estimate it from the residuals (e_t) estimated from the original model and then estimate the regression:

$$e_t = \hat{\rho} e_{t-1} + v_t \qquad (5.34)$$

$\hat{\rho}$ is a consistent estimate of ρ. We can use the estimated ρ to obtain $\hat{V}$, the estimated V, and use the EGLS to estimate the parameters of the LRM under AR(1).

This procedure can be generalized to the AR(2), AR(3), . . . , AR(p) correlation structure. Several statistical packages do this job fairly quickly.

5.6.1 The Newey–West Method of Correcting OLS Standard Errors[13]

Instead of speculating on the nature of autocorrelation or trying various methods to correct for it, why not find a procedure to correct for the OLS standard errors in cases of autocorrelation? This is precisely what Newey and West have done in their **HAC (heteroscedasticity- and autocorrelation-consistent) standard errors** or simply **Newey–West standard errors**. Although the mathematics behind HAC standard error is involved, the HAC method is now incorporated in several statistical packages. It may be noted that White's standard errors correct only for heteroscedasticity, whereas HAC corrects for both heteroscedasticity and autocorrelation. Like White's heteroscedasticity-corrected standard errors, HAC standard errors are also robust. The importance of robust standard errors lies in the fact that "in large samples they provide accurate hypothesis tests and confidence intervals given minimal assumptions about the data and model."[14]

---

[13]Newey, W. K., & West, K. (1987). A simple positive semi-definite heteroscedasticity and autocorrelation consistent covariance matrix. *Econometrica, 55,* 703–708.

[14]Angrist, J. D., & Pischke, J.-S. (2009). *Mostly harmless econometrics: An empiricist's companion* (p. 45). Princeton, NJ: Princeton University Press.

5.7 Summary

A critical assumption of the CLRM $y = BX + u$ is that $\text{cov}(uu') = \sigma^2 I$, that is, the error variance is homoscedastic and there is no autocorrelation among the error terms. Such a covariance matrix is called a **scalar** covariance matrix. But in many applications, such an assumption may not be appropriate. In cross-sectional data, for example, heteroscedasticity is frequently a problem. Similarly, in time-series data, autocorrelation is often encountered. In panel data, consisting of time-series and cross-sectional observations, we may have both these problems.

To deal with the problems of heteroscedasticity and autocorrelation, we considered a covariance matrix of the form $\text{cov}(uu') = \sigma^2 V$, where the matrix V is not necessarily an identity matrix, thus allowing for heteroscedasticity and autocorrelation in the error term. This is the case of the **nonscalar** covariance matrix.

If we routinely estimate a regression model assuming that the covariance matrix is scalar, whereas in fact it may not be, the OLS estimators are still unbiased as well as consistent, but they are no longer BLUE. The method of GLS is developed to handle a nonscalar covariance matrix. There are several diagnostic tests to find out whether in a concrete application we have a nonscalar covariance matrix.

To estimate the LRM with a nonscalar covariance matrix, we can proceed in two ways. Since V is a PD matrix, we can find a lower triangular matrix P, which we can use to transform the original model in such a way that the transformed model satisfies the OLS assumptions. This is basically the idea behind GLS: *GLS is OLS applied to the transformed model*. The transformed model produces estimators that are BLUE.

Instead of transforming the model as just suggested, and assuming that the regression error term is normally distributed with zero mean and constant variance, we can use the method of ML to estimate the parameters of the LRM with a nonscalar covariance matrix. Both of these approaches yield identical results. But note that the ML estimators have desirable properties only in *large samples*.

The main problem in the application of the GLS is that the covariance matrix V is generally not known. We can estimate V from the OLS residuals and use it to obtain GLS estimators. The estimated V is a consistent estimator of the unknown V. If we use the estimated V, the GLS is known as EGLS, estimated GLS.

Instead of estimating V by one or more methods, we can use White's heteroscedasticity-consistent standard errors to correct the heteroscedasticity problem. Or we can use Newey–West HAC standard errors, which take care of both heteroscedasticity and autocorrelation, depending on the situation. However, both of these methods for correcting heteroscedasticity and autocorrelation require large samples.

Exercises

5.1 Consider the following model:[15]

$$Y_i = BX_i + u_i, \quad i = 1, 2$$

where $u_1 \sim N(0, \sigma^2)$ and $u_2 \sim N(0, 2\sigma^2)$, and they are statistically independent. If $X_1 = +1$ and $X_2 = -1$, obtain

    a. the weighted least-squares estimate of B.

    b. the variance of the estimate.

5.2 Let $Z = c + Dy$, where y is a random vector, D is a fixed matrix, and c is a fixed vector. Show that

    a. $E(Z) = c + DE(y)$

and

    b. $\operatorname{cov}(Z) = D\Sigma_{yy} D'$

where Σ_{yy} is the covariance of y.

5.3 Consider the following model without the intercept:

$$Y_i = BX_i + u_i$$

You are told that $\operatorname{var}(u_i) = \sigma^2 X_i^2$. Show that

$$\operatorname{var}(B) = \frac{\sigma^2 \Sigma X_i^4}{(\Sigma X_i^2)^2}$$

5.4 Which of the following statements are true or false?

    a. In regression analysis, the assumption that the error term u is normally distributed is essential to validate the use of F and t tests.

    b. $H_0: B_1 = B_2 B_3$ is not a linear hypothesis.

    c. The formula $E(b) = B + (X'X)^{-1} X'X_2 B_2$ gives the bias effect on b of failing to include the terms $X_2 B_2$ in the model.

5.5 Suppose in the hypothetical savings–income regression discussed in the text, the error u_i has the following error variance structure:

$$E(u_i^2) = \sigma^2 X_i$$

---

[15]See Seber, G. A. F. (1977). *Linear regression analysis* (p. 64). New York, NY: Wiley (notations altered).

That is, the error variance is proportional to the level of income X. How would you transform the savings–income regression so that in the transformed model the error variance is homoscedastic?

Appendix 5A: ML Estimation of GLS

Consider the GLM

$$y = XB + u \tag{5A.1}$$

Now assume

$$\text{cov}(u) = E(uu') = V \tag{5A.2}$$

For simplicity, we have dropped the scalar σ^2 in front of V, where V is a PD matrix.

Assume that u follows a multivariate normal distribution. The joint density of $u_1, u_2, \ldots, u_n$ is therefore

$$f(u; \beta, V) = \frac{1}{(2\pi)^{n/2} |V|^{1/2}} \exp\left[-\frac{1}{2} u' V^{-1} u\right] \tag{5A.3}$$

Since V is PD, V^{-1} is also PD, and so is $u' V^{-1} u$. Therefore, to maximize (5A.3) we have to minimize $u' V^{-1} u$. We thus have to find b^{gls} that minimizes

$$u' V^{-1} u = (y - Xb^{\text{gls}})' V^{-1} (y - Xb^{\text{gls}}) \tag{5A.4}$$

You will observe a similarity with the ML estimator of the classical LRM. Call the left-hand side of (5A.4) the residual sum of squares, RSS_V.

Differentiate (5A.4) with respect to b^{gls} to yield

$$\frac{\text{RSS}_V}{\partial b^{\text{gls}}} = \frac{\partial(y - Xb^{\text{gls}})' V^{-1} (y - Xb^{\text{gls}})}{\partial b^{\text{gls}}}$$

$$= \frac{\partial}{\partial b^{\text{gls}}} \left[yV^{-1}y - 2b^{\text{gls}} X' V^{-1} y + b^{\text{gls}} X' V^{-1} Xb^{\text{gls}} \right]$$

$$= -2X' V^{-1} y + 2X' V^{-1} Xb^{\text{gls}} \tag{5A.5}$$

Setting the preceding equation to zero, we obtain

$$X'V^{-1}Xb^{gls} = X'V^{-1}y \tag{5A.6}$$

This gives

$$b^{gls} = (X'V^{-1}X)^{-1}X'V^{-1}y \tag{5A.7}$$

which is precisely the result obtained in (5.13), using the P transformation. The covariance matrix of b^{gls} is given in (5.14).

CHAPTER 6. EXTENSIONS OF THE CLASSICAL LINEAR REGRESSION MODEL: THE CASE OF STOCHASTIC OR ENDOGENOUS REGRESSORS

6.1 Introduction

Our discussion of the LRM, classical or normal classical, has been based on the assumption that the regressor values are fixed in repeated sampling. In the case where the regressors are random, we assumed that our analysis is *conditional* on the given values of regressors, which is clear from the discussion in Chapter 3. This is also clear from the assumption that

$$E(u\,|\,X) = 0 \qquad (6.1)$$

This assumption rules out any correlation between the error term and one or more regressors. In many applications, this assumption may be too restrictive or untenable. There are several reasons why the error term and one or more regressors are correlated, such as the following:[1]

1. Measurement errors in the regressor(s)

2. Omitted variable bias

3. Simultaneous equation bias

4. Dynamic regression models with serial correlation in the error term

So far, we have treated the regressand as random, but assumed the regressors take fixed values in repeated sampling. But now, we allow the regressors to be random or stochastic as well. We call this the **stochastic regressor case**, in contrast to the **fixed regressor case**. More specifically, we draw n independent observations from the multivariate population:

$$(y_1, x_1'), \ldots, (y_n, x_n') \qquad (6.2)$$

where x_i' denotes the ith row of the data matrix X. The rows of the data matrix (y_i, x_i') are assumed to be iid. Written more fully, (6.2) becomes

---

[1]For detailed reasons and illustrations, see Gujarati, D. (2015). *Econometrics by example* (2nd ed., chap. 19). London, England: Palgrave Macmillan.

$$Z = \begin{bmatrix} Y_1 & X_{11} & X_{21} & \cdots & X_{k1} \\ Y_2 & X_{12} & X_{22} & \cdots & X_{k2} \\ \vdots & & \ddots & & \vdots \\ \vdots & & & \ddots & \vdots \\ Y_n & X_{1n} & X_{2n} & \cdots & X_{kn} \end{bmatrix}$$

We call Z the stochastic regressor matrix.

We now consider three cases:

1. X and u are distributed independently.

2. X and u are contemporaneously (i.e., at the same time) uncorrelated.

3. X and u are neither independently distributed nor contemporaneously uncorrelated.

We consider each case separately.

6.2 X and u Are Distributed Independently

In this situation, the following holds true.

If the stochastic regressor matrix Z is independent of the random equation error vector e, then the least-squares estimator b, which is identical to the ML estimator, is unbiased. The estimator of the variance of the equation error, $\hat{\sigma}^2 = \dfrac{\hat{e}'\hat{e}}{T-K}$, is an unbiased estimator of σ^2, whereas $\tilde{\sigma}^2 = \dfrac{\hat{e}'\hat{e}}{T}$ is a biased estimator of σ^2. The independence of the equation error and the stochastic regressors ensures that these estimators have the classical properties conditional for any matrix values of Z, and classical interval estimation and test procedures remain valid.[2]

Note that the notations Z, e, $\hat{\sigma}^2$, and T correspond to our X, u, S^2, and n.

It may be added that the probability distribution of X, since it is random now, does not involve the OLS parameters B and σ^2.

---

[2]This follows Judge, G. G., Carter Hill, R., Griffiths, W. E., Lutkepohl, H., & Lee, T.-C. (1988). *Introduction to the theory and practice of econometrics* (2nd ed., chap. 13, p. 574). Hoboken, NJ: Wiley. See also Goldberger, A. S. (1991). *A course in econometrics* (chap. 25). Cambridge, MA: Harvard University Press, for a slightly different treatment of this subject.

6.3 X and u Are Contemporaneously Uncorrelated

This is a weaker condition than the first one. In this case, the classical *results hold only asymptotically*, that is, in large samples. This is discussed further in Appendix 6B.

6.4 X and u Are Neither Independently Distributed Nor Contemporaneously Uncorrelated

In this, the more serious case, the *OLS estimators are not only biased but also inconsistent*. We show this first with the simple case of the bivariate regression:

$$Y_i = B_1 + B_2 X_i + u_i \tag{6.3}$$

The OLS estimation of B_2 in (6.3) is

$$b_2 = \frac{\Sigma x_i y_i}{\Sigma x_i^2} = \frac{\Sigma x_i Y_i}{\Sigma x_i^2} \tag{6.4}$$

where $x_i = (X_i - \bar{X})$ and $y_i = (Y_i - \bar{Y})$.

Now substituting Equation (6.3) into the right-hand side of Equation (6.4), we obtain

$$
\begin{aligned}
b_2 &= \frac{\Sigma x_i (B_1 + B_2 X_i + u_i)}{\Sigma x_i^2} \\
&= B_1 \frac{\Sigma x_i}{\Sigma x_i^2} + B_2 \frac{\Sigma x_i X_i}{\Sigma x_i^2} + \frac{\Sigma x_i u_i}{\Sigma x_i^2} \\
&= B_2 + \frac{\Sigma x_i u_i}{\Sigma x_i^2}
\end{aligned}
\tag{6.5}
$$

where we have used the fact that $\Sigma x_i = 0$, because the sum of deviations of a random variable from its mean value is always zero, and also because $\frac{\Sigma x_i X_i}{\Sigma x_i^2} = 1$. $(Hint: \Sigma x_i^2 = X_i^2 - n\bar{X}^2)$

Now if we take the expectation of the last term on the right-hand side of Equation (6.5), we run into a problem, for

$$E\left(\frac{\Sigma x_i u_i}{\Sigma x_i^2}\right) \neq \frac{E(\Sigma x_i u_i)}{E(\Sigma x_i^2)} \tag{6.6}$$

because the expectation operator, E, is a linear operator. Furthermore, the expectation of the product of x_i and u_i is not the product of the expectations, because they are not independent.[3]

The best we can do is see the behavior of b_2 as the sample size increases indefinitely. For this purpose, we use the concept of **probability limit**, or *plim* for short, which is the standard procedure to find out if an estimator is consistent, that is, if it approaches the true (population) value as the sample size increases indefinitely (see Appendix B on the properties of *plim*). So we proceed as follows:

$$
\begin{aligned}
p\lim(b_2) &= p\lim\left(B_2 + \frac{\Sigma x_i u_i}{\Sigma x_i^2}\right) \\
&= B_2 + p\lim\left(\frac{\Sigma x_i u_i}{\Sigma x_i^2}\right) \\
&= B_2 + \frac{p\lim(\Sigma x_i u_i / n)}{p\lim(\Sigma x_i^2 / n)} \\
&= B_2 + \frac{\text{Population cov}(X_i, u_i)}{\text{Population var}(X_i)}
\end{aligned}
\tag{6.7}
$$

where use is made of the properties of *plim*,[4] n is the sample size, and "cov" means covariance and "var" means variance.

As a result, we obtain

$$
p\lim(b_2) - B_2 = \frac{\text{cov}(X_i, u_i)}{\text{var}(X_i)}
\tag{6.8}
$$

which may be called the asymptotic bias.

Now, if the covariance between the regressor and the error term is positive, b_2 will overestimate the true B_2, a positive bias. On the other hand, if the covariance term is negative, b_2 will underestimate B_2, a negative bias. Moreover, the bias, positive or negative, will not disappear no matter how large the sample size is.

In summary, if a regressor and the error term are correlated, the OLS estimator is biased as well as inconsistent. Even if a single regressor in a multiple regression is correlated with the error term, the OLS estimators of

---

[3]Remember that $E(XY) = E(X)E(Y)$ only if X and Y are independent.

[4]In brief, these properties are as follows: $p\lim(X + Y) = p\lim X + p\lim Y$; $p\lim(XY) = p\lim X \cdot p\lim Y$; $p\lim(X/Y) = p\lim X / p\lim Y$; and the $p\lim$ of a constant is that constant itself.

all the coefficients are inconsistent.[5] In consequence, the traditional tests of hypotheses based on the t and F tests are invalid.

6.5 The Case of k Regressors[6]

Before we consider remedies for the problem we just discussed, we consider the general case of k regressors, $y = XB + u$, where the data matrix X is stochastic, that is, one or more regressors in it are correlated with the error term.

We have already shown that

$$b = B + (X'X)^{-1}X'u \quad \text{(same as Equation 2.7)} \tag{6.9}$$

and that

$$E(b) = B + E[(X'X)^{-1}X'u] \tag{6.10}$$

If X and u are independently distributed, we obtain

$$\begin{aligned} E(b) &= B + E\,[(X'X)^{-1}X']E(u) \\ &= B \end{aligned} \tag{6.11}$$

knowing that the error term has zero mean value.

However, if X and u are not independent, we cannot take the expectation of (6.10), and therefore, b will be biased and it may also not be consistent. To see if it is consistent, we proceed as follows:

Let

$$Q = p\lim(n^{-1}X'X) \tag{6.12}$$

where Q is a PD nonsingular matrix, and

$$p\lim(n^{-1}X'u) \neq 0 \tag{6.13}$$

that is, in the limit the errors are correlated with at least one regressor.

---

[5]Recall that in multiple regression, the cross-product terms of the regressors are involved in the computation of the partial regression coefficients. Therefore, an error in a regressor may affect the coefficients of the other regressors in the model.

[6]This is based on the discussion in Darnell, A. C. (1994). *A dictionary of econometrics* (pp. 197–200). Cheltenham, England: Edward Elgar; Paul de Boer, C. H., Franses, P. H., Kloek, T., & van Dijk, H. K. (2004). *Econometric methods with applications in business and economics* (chap. 5). Oxford, England: Oxford University Press.

Taking the $p$lim of (6.9), we obtain

$$\text{plim}(\boldsymbol{b}) = \boldsymbol{B} + \text{plim}(n^{-1}X'X)^{-1}\text{plim}(n^{-1}X'\boldsymbol{u})$$
$$= \boldsymbol{B} + \boldsymbol{Q}^{-1}\text{plim}(n^{-1}X'\boldsymbol{u})$$
$$\neq \boldsymbol{B} \tag{6.14}$$

This shows that the OLS estimators are inconsistent.

Now consider the covariance matrix of $\boldsymbol{b}$:[7]

$$\text{cov}(\boldsymbol{b}) = E\left[(\boldsymbol{b}-\boldsymbol{B})(\boldsymbol{b}-\boldsymbol{B})'\right] = E\left[(X'X)^{-1}X'\boldsymbol{u}\boldsymbol{u}'X(X'X)^{-1}\right]$$
$$= E\left\{(X'X)^{-1}X'\left[E(\boldsymbol{u}\boldsymbol{u}' \mid XX)\right]X(X'X)^{-1}\right\} \tag{6.15}$$
$$= \sigma^2 E(X'X)^{-1}$$

This covariance matrix is different from the classical formula $\sigma^2(X'X)^{-1}$.

As a result, the usual tests of hypotheses are not valid and the standard t test does not have the t distribution. The test procedure in this case depends on the probability distribution of X. Additionally, the OLS estimator $\boldsymbol{b} = (X'X)^{-1}X'y$ is now a stochastic function of y and is no longer BLUE. However, the estimator is efficient if we assume that our analysis is conditional on the given X.

But in the limit if $\boldsymbol{u}$ is uncorrelated with all the regressors, that is, $\text{plim}(n^{-1}X'\boldsymbol{u}) = 0$, then $\boldsymbol{b}$ is a consistent estimator of $\boldsymbol{B}$. Also, $\boldsymbol{b}$ is an ML estimator if $\boldsymbol{u}$ is normally distributed and the probability distribution of X does not involve the OLS parameters, as noted before.

6.6 What Is the Solution? The Method of Instrumental Variables (IVs)

As we saw in the case of stochastic regressors, the OLS estimators are not consistent because of the correlation between the regressor(s) and the error term. Can we find a set of regressors that are highly correlated with the original regressors and yet are uncorrelated with the error term? If we can find such variables, called **instrumental variables (IVs)** or **proxy**

---

[7]In deriving this, note that "if $f(x, y)$ is the joint density function of the random variables X and Y, and if $h(x, y)$ is function of X and Y such that $E[h(X, Y)]$ exists, then $E\left[h(X,Y)\right] = E_X\left\{E_{Y|X}\left[h(X,Y)\right]\right\}$, where $E_{Y|X}$ is the expectation in the conditional distribution of Y given X and E_X is the marginal distribution of X." Schmidt, P. (1976). *Econometrics* (pp. 94–95). New York, NY: Marcel Dekker.

variables, a big if, we can obtain consistent estimators of B. They are called "instrumental variables essentially because they lead to a transformation of the original model that is *instrumental* to solving the current inverse problem in which the orthogonality between X and $\varepsilon[=u]$ does not hold."[8]

Let Z be an $(n \times k)$ matrix such that

$$p\lim(n^{-1}Z'X) = \Sigma_{ZX}, \quad \text{rank of } \Sigma_{ZX} = k \tag{6.16}$$

where Z is a finite nonsingular $(k \times k)$ matrix and

$$p\lim(n^{-1}Z'u) = 0 \tag{6.17}$$

Note: Contrast (6.17) with (6.13).

$$p\lim(n^{-1}Z'Z) = \Sigma_{ZZ}, \quad \text{a nonsingular matrix of rank } k \tag{6.18}$$

The k variables included in Z are in the limit uncorrelated with the error term but are correlated with the k variables in X. The variables in Z are called IVs or simply the instruments. The (chosen) instruments must satisfy the following conditions:

1. *Instrument relevance*: That is, Z must be correlated with the stochastic variables for which they act as an instrument, X in our case. The greater the extent of correlation between Z and X, the better are the instruments. Symbolically,

$$\text{cov}(X, Z) \neq 0. \tag{6.19}$$

2. *Instrument exogeneity*: Z must not be correlated with u, that is,

$$\text{cov}(Z, u) = 0 \tag{6.20}$$

3. *Variables included in Z are not regressors in their own right:* That is, they do not belong in the original model. If they do, the original model must be misspecified.

4. There must be at least as many IVs as the number of stochastic regressors in the model.

---

[8]Mittelhammer, R. C., Judge, G. G., & Miller, D. J. (2000). *Econometric foundations* (p. 424). Cambridge, England: Cambridge University Press.

There are several questions about these conditions, which we will answer after we show how the IV method works.

But at this stage, it is worth mentioning the following:

1. If the number of instruments (I) equals the number of endogenous regressors, say, k, we say that the regression coefficients are **exactly identified**, that is, we can obtain unique estimates of them. In this chapter, we only consider this case; the theoretical discussion of the two cases mentioned below is beyond the scope of this book, although the examples discussed later in the chapter shed some light on them.

2. If the number of instruments (I) exceeds the number of endogenous regressors, k, the regression coefficients are **overidentified**, in which case we may obtain more than one estimate of coefficients of one or more regressors.

3. If the number of instruments is less than the number of endogenous regressors, the regression coefficients are **underidentified**, that is, we cannot obtain unique values of the regression coefficients.[9]

6.6.1 IV Regression

Now consider the following estimator of B, called the IV estimator:

$$b^{IV} = (Z'X)^{-1} Z' y \tag{6.21}$$

Notice the similarity with the OLS estimator $b = (X'X)^{-1} X' y$.

Substituting for $y = XB + u$, we obtain

$$b^{IV} = (Z'X)^{-1} Z'(XB + u)$$
$$= B + (Z'X)^{-1} Z' u \tag{6.22}$$

Taking $p$lims of (6.22), we obtain

$$p\text{lim}(b^{IV}) = B + p\text{lim}[(n^{-1}Z'X)^{-1}] p\text{lim}[n^{-1}Z'u]$$
$$= B + \Sigma_{ZX}^{-1} x \cdot 0$$
$$= B \tag{6.23}$$

where use is made of (6.16) and (6.17) and the properties of $p$lim.

---

[9]This topic is usually discussed in the context of simultaneous equation models. See Gujarati, D. N., & Porter, D. C. (2009). *Basic econometrics* (5th ed., chaps. 18, 19, and 20). New York, NY: McGraw-Hill.

As (6.23) shows, the IV estimator b^{IV} is a consistent estimator of B, and the k variables in Z are called IVs or simply the instruments. Interestingly, OLS is a special case of IV if the regressors in X are uncorrelated with u in the limit. In this case, X serves as its own instruments.

6.6.2 Estimation of the IV Regression Error Variance

Let $e_{IV} = y - Xb^{IV}$ denote the IV residuals. The IV error variance is defined as

$$S_{IV}^2 = \frac{1}{n-k}(y - Xb^{IV})'(y - Xb^{IV}) \tag{6.24}$$

Compare this with the error variance of OLS given in (2.18).

6.6.3 Covariance Matrix of IV Estimators

The asymptotic covariance matrix of b^{IV} can be derived as

$$\text{cov}(b^{IV}) = \frac{\sigma^2}{n}\left[\Sigma_{ZX}^{-1}\Sigma_{ZZ}\Sigma_{XZ}^{-1}\right] \tag{6.25}$$

In practice, we estimate this covariance matrix as follows:

$$\text{cov}(b^{IV}) = S_{IV}^2(Z'X)^{-1}(ZZ')(X'Z)^{-1} \tag{6.26}$$

It can be shown that asymptotically,

$$b^{IV^{asy}} \sim N\left[B, \frac{\sigma^2}{n}(\Sigma_{ZX}^{-1}\Sigma_{ZZ}\Sigma_{XZ}^{-1})\right] \tag{6.27}^{[10]}$$

The IV estimators, although consistent in large samples, are biased in small samples. Moreover, they are not necessarily more efficient than the OLS estimators, but they are relatively more efficient if the instruments Z are highly correlated with the X variables.

To show this, we consider the simple case of bivariate regression:

$$Y_i = B_1 + B_2 X_i + u_i \tag{6.28}$$

---

[10]See Greene, W. H. (2012). *Econometric analysis* (7th ed., pp. 226–227). New York, NY: Prentice Hall.

Assume that X_i and u_i are correlated, and we "find" an IV Z_i, which while correlated with X_i is uncorrelated with u_i.

Letting $x_i = (X_i - \bar{X})$, $y_i = (Y_i - \bar{Y})$, $z_i = (Z_i - \bar{Z})$, it is easy to show that

$$b_2 = \frac{\Sigma x_i y_i}{\Sigma x_i^2} \tag{6.29}$$

$$b_2^{IV} = \frac{\Sigma z_i y_i}{\Sigma z_i x_i} \tag{6.30}$$

Using (6.28), we can write

$$y_i = B_2 x_i + (u_i - \bar{u}) \tag{6.31}$$

where $\bar{u}$ is the average value population error term, $\boldsymbol{u}$.

Substituting for y_i from (6.31), into (6.30), we obtain

$$
\begin{aligned}
b^{IV} &= \frac{\Sigma z_i [B_2 x_i + (u_i - \bar{u})]}{\Sigma z_i x_i} \\
&= B_2 + \frac{\Sigma z_i (u_i - \bar{u})}{\Sigma z_i x_i}
\end{aligned}
\tag{6.32}
$$

Since the population $\text{cov}(\boldsymbol{Z}, \boldsymbol{u}) = 0$ by assumption, taking the $p\lim$ of (6.32), we obtain

$$p\lim b_2^{IV} = B_2 \tag{6.33}$$

that is, b_2^{IV} is a consistent estimator of B_2.

Specializing (6.26) to the bivariate regression, it can be shown that in large samples the IV estimator for the bivariate regression model is distributed as follows:

$$b_2^{IV} \sim N\left(B_2, \frac{\sigma^2}{\Sigma x_i^2} \frac{1}{\rho_{XZ}^2}\right) \tag{6.34}$$

where ρ_{XZ} is the (population) correlation coefficient between X and Z. If ρ is 1, then we are back to the standard OLS estimator, for in this case X acts as its own instrument. As you can see from (6.34), if the correlation coefficient is rather low, the variance of the IV estimator will be greater than the OLS variance. In this case, the chosen Z instrument is a *weak instrument*. Thus, there is a trade-off between consistency and efficiency. *In fact, the efficiency of the IV estimator depends on the degree of correlation between the instruments and the* **X** *regressors.*

6.7 Hypothesis Testing Under IV Estimation

Can we use the traditional t, F, and χ^2 to test hypotheses under IV estimation? In finite samples, we cannot use these tests. As Davidson and MacKinnon note, "Because the finite sample distributions of IV estimators are almost never known, exact tests of hypotheses based on such estimators are almost never known."[11]

However, as we have shown, in large samples the IV estimators are approximately normally distributed with mean and variance given previously. Therefore, instead of using the standard t test, we use the Z test (i.e., the standard normal distribution). To test joint hypotheses about two or more IV coefficients, instead of the classical F test, we use the large-sample *Wald test*. As we have discussed earlier, the Wald statistic follows the χ^2 statistic with degrees of freedom equal to the number of regressors estimated.

Some computer packages publish R^2 for the IV regression. But it does not have the same meaning as in the standard linear regression, and sometimes it can actually be negative.[12]

6.8 Practical Problems in the Application of the IV Method

In using the IV method, we need to answer some important questions:

1. How do we know that a regressor is truly stochastic or endogenous?

2. How do we find out if an instrument is weak or strong?

3. How do we test the validity of all the instruments?

4. How do we estimate a model when there is more than one stochastic regressor? Remember that the number of instruments must be at least as great as the number of endogenous regressors.

5. What happens if there are more instruments than the number of endogenous regressors?

We answer these questions below.

---

[11]Davidson, R., & MacKinnon, J. G. (2004). *Econometric theory and methods* (2nd ed., pp. 330–335). New York, NY: Oxford University Press.

[12]The traditional R^2 is defined as $R^2 = 1 - \text{RSS/TSS}$, but in the case of IV estimation, RSS can be greater than TSS, making R^2 negative.

6.8.1 Test of the Endogeneity of a Regressor

To test whether a variable is endogenous, we can use one of the variants of the **Hausman test**.[13] The test involves two steps:

Step 1: Regress the suspected endogenous variable on all the nonstochastic regressors in the model plus the IV(s) and obtain residuals from this regression, call it Rs_1.

Step 2: Regress the dependent variable on all the regressors, including the endogenous ones, and the residual Rs_1 obtained in Step 1. If the t (or p value) of the residual Rs_1 in this regression is statistically significant, we conclude that the suspected endogenous variable is in fact endogenous. If it is not, then there is no need for the IV estimation.

6.8.2 How to Find Whether an Instrument Is Weak or Strong

If the instrument used in the analysis is weak, in the sense that it is poorly correlated with the endogenous regressor for which it is a proxy, the IV estimator can be severely biased and its sampling distribution is not approximately normal even in large samples. Thus, the standard errors and the confidence intervals based on them are highly misleading, leading to hypothesis tests that are unreliable.

How do we decide in a given situation whether an instrument is weak? In the case of a single endogenous regressor, a rule of thumb says that an F statistic of less than 10 in the first step of the Hausman test suggests that the chosen instrument is weak. If it is greater than 10, it probably is not a weak instrument.[14] In the case of a single stochastic regressor, this rule translates into a t value of about 3.2 because of the relationship between the F and t statistics: $F_{1,k} = t_k^2$, where the F statistic has 1 df in the numerator and k df in the denominator. Of course, this rule of thumb, like most rules of thumb, should be used judiciously.

6.8.3 The Case of Multiple Instruments

Since there may be competing instruments for an endogenous regressor, it may be correlated with more than one IV. To allow for this possibility, we

---

[13]Hausman, J. (1978). Specification tests in econometrics. *Econometrica, 46,* 1251–1271.

[14]Why 10? The technical answer to this can be found in Stock, J. H., & Watson, M. W. (2007). *Introduction to econometrics* (2nd ed., p. 466). Boston, MA: Pearson/Addison. If the F statistic exceeds 10, it suggests that the small sample bias of the IV estimate is less than 10% of the OLS bias. Remember that in cases of stochastic regressors, OLS estimators are biased in small as well as in large samples.

can include more than one instrument in the IV regression. This is often done with the aid of **two-stage least squares (2SLS)**.

Step 1: Regress the suspected endogenous variable on all the instruments and obtain its estimated value from this regression.

Step 2: Run the regression you want to estimate on all the regressors included in the model of interest and replace the endogenous regressor by its value estimated in Step 1.

We can replace this two-step procedure by a single step by invoking *Stata*'s *ivreg* command by including several instruments simultaneously.

6.8.4 Testing the Validity of Surplus Instruments[15]

The steps involved in this test are as follows:

1. Regress the dependent variable on all the exogenous variables in the model plus all the instruments.

2. Obtain residuals from this regression; call them *Res*.

3. Regress *Res* on all the original regressors, including the instruments, and obtain an R^2 value from this regression.

4. Multiply this R^2 value by the number of observations in the sample, n, and compute nR^2. If all the surplus instruments are valid, it can be shown that

$$nR^2 \sim \chi_m^2$$

where m is the number of surplus instruments.

5. If the estimated chi-square value exceeds the critical chi-square value, say at the 5% level, conclude that at least one surplus instrument is *not* valid.

6.9 Regression Involving More Than One Endogenous Regressor

Just as one instrument was sufficient to identify the impact of an endogenous regressor, we need two instruments to identify the impact of two endogenous regressors, and so on. The steps involved in this case are as follows:

---

[15]The following discussion is based on Carter Hill, R., Griffiths, W. E., & Lim, G. C. (2008). *Principles of econometrics* (3rd ed., pp. 289–290). New York, NY: Wiley.

Stage 1: Regress each endogenous regressor on all the exogenous variables and obtain the estimated values of these regressors.

Stage 2: Regress the dependent variable on all the exogenous variables and the estimated values of the endogenous regressors from Stage 1.

We can bypass the two-step procedure; packages such as *Stata* can do this in one step.

We have covered a lot of material on IV estimation. We now illustrate the various points made with a concrete example.

6.10 An Illustrative Example: Earnings and Educational Attainment of Youth in the United States

The National Longitudinal Survey of Youth 1979 (NLSY79) is a repeated survey of a nationally representative sample of young males and females between ages 14 and 21 years in 1979. From 1979 until 1994, the survey was conducted annually, but since then, it has been conducted biannually. Originally, the core sample consisted of 3,003 males and 3,108 females.

The NLSY cross-sectional data are provided in 22 subsets, each subset consisting of a randomly drawn sample of 540 observations; 270 males and 270 females.[16] Data are collected on a variety of socioeconomic conditions and are quite extensive. The major categories of data obtained pertain to gender, ethnicity, age, years of schooling, highest educational qualification, marital status, faith, family background (mother's and father's education and number of siblings), place of living, earnings, hours, years of work experience, type of employment (government, private sector northeastern, self-employed), and region of the country (north central, northeastern, southern, and western). (The full data set is available on the book's website https://study.sagepub.com/gujarati)

We will use some of these data for 2002 (sample subset number 22) to develop an earnings function. Following the tradition established by Jacob Mincer, we consider the following earnings function:[17]

---

[16]The data used here are obtained from the website of Dougherty, C. (2007). *Introduction to econometrics* (3rd ed.). New York, NY: Oxford University Press. The data can be obtained directly from www.bls.gov/nls. Some of the data can be downloaded, and more extensive data can be purchased. The Dougherty text also has a good discussion of instrumental variables in chap. 8.

[17]Mincer, J. (1974). *Schooling, experience, and earnings.* New York, NY: Columbia University Press. See also Hickman, J. J., Lochner, L. J., & Todd, P. E. (2003, May). *Fifty years of Mincer earnings functions* (Working Paper No. 9732). Cambridge, MA: National Bureau of Economic Research. On functional forms of regression models, see Chapter 7.

$$\ln Earn = B_1 + B_2 S + B_3 Wexp + B_4 Gender + B_5 Ethblack + B_6 Ethhisp + u_t \quad (6.35)$$

where $\ln Earn$ = log of hourly earnings in dollars, S = years of schooling (highest grade completed in 2002), $Wexp$ = total out-of-school work experience in years as of the 2002 interview, $Gender = 1$ for female and 0 for men, $Ethblack = 1$ for blacks, $Ethhisp = 1$ for Hispanic; nonblack and non-Hispanic, which are reference categories, are left out.

As you can see, some variables are quantitative and some are dummy variables. A priori, and based on prior empirical evidence, we expect

$$B_2 > 0, B_3 > 0, B_4 < 0, B_5 < 0, \text{and } B_6 < 0.$$

For the purpose of this chapter, our concern is with the education variable S in the above model. If (native) ability and education are correlated, we should include both variables in the model. However, the ability variable is difficult to measure directly. As a result, it may be subsumed in the error term. But in that case the education variable may be correlated with the error term, thereby making education an endogenous or a stochastic regressor. From our discussion of the consequences of stochastic regressor(s), it would seem that if we estimate Equation (6.35) by OLS, the coefficient of S will be biased as well as inconsistent. This is so because we may not be able to find the true impact of education on earnings that does not net out the effect of ability. Naturally, we would like to find a suitable instrument or instruments for years of schooling so that we can obtain a consistent estimate of its coefficient.

Before we search for the instrument(s), let us estimate Equation (6.35) by OLS for comparative purposes. The regression results using *Stata 10* are shown in Table 6.1.

All estimated coefficients have the expected signs, and under the classical assumptions all coefficients are statistically highly significant, the sole exception being the dummy coefficient for Hispanics.

These results show that compared with male workers, female workers on average earn less than their male counterparts, ceteris paribus. The average hourly earnings of black workers is lower than that of nonblack non-Hispanic workers, ceteris paribus, which is the base category. Qualitatively, the sign of the Hispanic coefficient is negative, but the coefficient is statistically insignificant.

Noting that the regression model is log-lin, we have to interpret the coefficients of quantitative and qualitative (i.e., dummy) variables carefully. For quantitative variables, schooling and work experience, the estimated coefficients represent **semielasticities**—that is, relative or percentage change in the dependent variable for a unit change in the value of the regressor, holding all other regressors constant. The results suggest that if

Table 6.1 Earnings Function, the United States, 2000 Data Set

```
regress lnearnings s female wexp ethblack ethhisp,robust
```

```
Linear regression                          Number of obs   =        540

                                           F(5, 534)       =      50.25

                                           Prob > F        =     0.0000

                                           R-squared       =     0.3633

                                           Root MSE        =     .50515
```

		Robust				
lnearnings	Coef.	Std. Err.	t	P>|t|	[95% Conf.	Interval]
s	.1263493	.0097476	12.96	0.000	.1072009	.1454976
female	-.3014132	.0442441	-6.81	0.000	-.3883269	-.2144994
wexp	.0327931	.0050435	6.50	0.000	.0228856	.0427005
ethblack	-.2060033	.062988	-3.27	0.001	-.3297381	-.0822686
ethhisp	-.0997888	.088881	-1.12	0.262	-.2743881	.0748105
_cons	.6843875	.1870832	3.66	0.000	.3168783	1.051897

Note: Regress is *Stata*'s command for OLS regression. This command is first followed by the dependent variables and then the regressors. Sometimes additional options are given, such as *robust*, which computes robust standard errors; in the present case, standard errors are corrected for heteroscedasticity, a topic we have discussed in Chapter 5.

schooling increases by a year, the average hourly earnings go up by about 13%, ceteris paribus. Similarly, if work experience goes up by 1 year, the average hourly earnings go up by about 3.2%, ceteris paribus.

To obtain the semielasticity of a dummy variable, we first take the antilog of the dummy coefficient, subtract 1 from it, and multiply the difference by 100%. Following this procedure, for the female dummy coefficient we obtain a value of about 0.7397, which is approximately the antilog of −0.3014, which suggests that females on average earn about 26% less than male workers. Following this procedure, the values for blacks and Hispanics are about 0.81 and 0.90, respectively. This suggests that black and Hispanic workers on average earn less than the base category by about 19%

and 10%, although the semielasticity for Hispanics is not statistically different from the base category.

As we have discussed, since the education variable does not necessarily take into account ability, it may be correlated with the error term, thus rendering it a stochastic regressor. If we can find a suitable instrument for schooling that satisfies the three requirements we specified for a suitable instrument, we can use it and estimate the earnings function by the IV method. The question is what may be a proper instrument? This question is difficult to answer categorically. What we can do is try one or more proxies and compare the OLS results given in Table 6.1 to see how far the OLS results are biased, if at all.

In the data we have information on mother's education (*Sm*), father's education (*Sf*), both measured by years of schooling, number of siblings, and the ASVAB verbal (word knowledge) and mathematics (arithmetic reasoning) scores.

In choosing a proxy or proxies, we must bear in mind that such proxies must be *uncorrelated* with the error term but must be correlated (presumably highly) with the stochastic regressor(s) and must *not* be candidates in their own right as regressors—in the latter case, the model used in the analysis will suffer from model specification errors. It is not always easy to accomplish all of these objectives in every case. So very often, it is a matter of trial and error, underlying economic theory, and prior empirical work, supplemented by judgment or "feel" for the participant under study.

However, there are diagnostic tests that can tell us if the chosen proxy or proxies are appropriate, tests which we will consider shortly. The data give information on mother's schooling (*Sm*), which we will use as the instrument for participant's schooling. The thinking here is that *S* and *Sm* are correlated—a reasonable assumption. For our data, the correlation between the two is about 0.40. We have to assume that *Sm* is uncorrelated with the error term. We also assume that *Sm* does not belong in the participant's earning function, which seems reasonable.

We accept for the time being the validity of *Sm* as an instrument, which we will test after we present the details of IV estimation.

To use *Sm* as the instrument for *S* and estimate the earnings function, we proceed in two stages:

Stage 1: We regress the suspected endogenous variable (*S*) on the chosen instrument (*Sm*) and the other regressors in the original model and obtain the estimated value of *S* from this regression; call it *sshat*.

Step 2: We then run the earnings regression on the regressors included in the original model but replace the education variable by its value estimated from the Step 1 regression, *sshat*.

This method of estimating the parameters of the model of interest is appropriately called the method of 2SLS, for we apply OLS twice. Therefore *the IV method is also known as 2SLS.*

Let us illustrate this method. Stage 1 regression is shown in Table 6.2. Using the estimated *S*-hat value from this regression, we obtain the second-stage regression 2SLS (Table 6.3).

Note that in this (log) earnings function we use *sshat* (estimated from the first stage of 2SLS) instead of *S* as the regressor. However, *the standard errors reported in Table 6.3 are not correct* because they are based on the incorrect estimator of the variance of the error term, u_i. The formula to correct the estimated standard errors is rather involved. So it is better to use software such as *Stata* or *Eviews*, which not only will correct the standard errors but also will obtain the 2SLS estimates without explicitly going through the cumbersome two-step procedure.

Table 6.2   First Stage of 2SLS With *Sm* as Instrument

```
Regress s female wexp ethblack ethhisp sm
```

Source	SS	df	MS	Number of obs	=	540
				F(5, 534)	=	35.06
Model	822.26493	5	164.452986	Prob > F	=	0.0000
Residual	2504.73322	534	4.69051164	R-squared	=	0.2471
				Adj R-squared	=	0.2401
Total	3326.99815	539	6.17253831	Root MSE	=	2.1658

s	Coef.	Std. Err.	t	P>|t|	[95% Conf. Interval]	
female	-.0276157	.1913033	-0.14	0.885	-.4034151	.3481837
wexp	-.1247765	.0203948	-6.12	0.000	-.1648403	-.0847127
ethblack	-.9180353	.2978136	-3.08	0.002	-1.503065	-.3330054
ethhisp	.4566623	.4464066	1.02	0.307	-.4202661	1.333591
sm	.3936096	.0378126	10.41	0.000	.3193298	.4678893
_cons	11.31124	.6172187	18.33	0.000	10.09876	12.52371

Table 6.3 Second Stage of 2SLS of the Earnings Function

```
regress lnearnings sshat female wexp ethblack ethhisp

    Source |       SS          df       MS         Number of obs  =      540
-----------+----------------------------------     F(5, 534)      =    24.26
     Model |  39.6153245        5    7.9230649      Prob > F       =   0.0000
  Residual |  174.395063      534    .326582515     R-squared      =   0.1851
-----------+----------------------------------     Adj R-squared  =   0.1775
     Total |  214.010387      539    .397050811     Root MSE       =   .57147


--------------------------------------------------------------------------------
lnearnings |    Coef.    Std. Err.      t      P>|t|     [95% Conf. Interval]
-----------+--------------------------------------------------------------------
     sshat |   .140068    .0253488     5.53    0.000     .0902724    .1898636
    female |  -.2997973   .0505153    -5.93    0.000    -.3990304   -.2005642
      wexp |   .0347099   .0064313     5.40    0.000     .0220762    .0473437
   ethblack|  -.1872501   .0851267    -2.20    0.028    -.3544744   -.0200258
   ethhisp |  -.0858509   .1146507    -0.75    0.454    -.3110726    .1393708
     _cons |   .4607715   .4257416     1.08    0.280    -.3755622    1.297105
```

To do this, we can use the **ivreg** (IV regression) command of *Stata*. Using this command, we obtain the results in Table 6.4.

Observe that the estimated coefficients in the preceding two tables are the same, but the standard errors are different. As pointed out, we should rely on the standard errors reported in Table 6.4. Also notice that with the *ivreg* command, we need only one table, instead of two, as in the case of the rote application of 2SLS.

6.10.1 Test of the Endogeneity of a Regressor

So far, we have been working on the assumption that S in our example is endogenous. But we can test this assumption explicitly by using one of the variants of the Hausman test. This test is relatively simple and involves two steps.

Table 6.4   One-Step Estimate of the Earnings Function (With Robust Standard Errors)

```
ivregress 2sls lnearnings female wexp ethblack ethhisp (s=sm),robust

Instrumental variables (2SLS) regression          Number of obs   =       540

                                                   Wald chi2(5)    =    138.45

                                                   Prob > chi2     =    0.0000

                                                   R-squared       =    0.3606

                                                   Root MSE        =    .50338


-------------------------------------------------------------------------------
             |               Robust
  lnearnings |    Coef.    Std. Err.      z     P>|z|    [95% Conf. Interval]
-------------+-----------------------------------------------------------------
           s |   .140068   .0217263    6.45    0.000    .0974852    .1826508
      female |  -.2997973   .043731   -6.86    0.000   -.3855085   -.2140861
        wexp |   .0347099  .0055105    6.30    0.000    .0239095    .0455103
     ethblack |  -.1872501  .0634787   -2.95    0.003   -.3116661   -.0628342
     ethhisp |  -.0858509  .0949229   -0.90    0.366   -.2718963    .1001945
       _cons |   .4607717  .3560759    1.29    0.196   -.2371242   1.158668
-------------------------------------------------------------------------------

Instrumented:  S (=Sm)

Instruments:   female wexp ethblack ethhisp sm
```

Step 1: We regress the endogenous S on all the (nonstochastic) regressors in the earnings function plus the IV(s) and obtain residuals from this regression; call it Res_1.

Step 2: We then regress ln Earnings on all the regressors, including the (stochastic) S and the residuals (Res_1) from Step 1. If in this regression the t value of the residuals variable is statistically significant, we conclude that S is endogenous. If it is not, then there is no need for IV estimation, for in that case S is its own instrument.

Returning to our example, we obtain the results shown in Tables 6.5 and 6.6.

Table 6.5 Hausman Test of Endogeneity of Schooling: First-Step Results

```
regress s female wexp ethblack ethhisp sm
```

Source	SS	df	MS			
				Number of obs	=	540
				F(5, 534)	=	35.06
Model	822.26493	5	164.452986	Prob > F	=	0.0000
Residual	2504.73322	534	4.69051164	R-squared	=	0.2471
				Adj R-squared	=	0.2401
Total	3326.99815	539	6.17253831	Root MSE	=	2.1658

s	Coef.	Std. Err.	t	P>|t|	[95% Conf.	Interval]
female	-.0276157	.1913033	-0.14	0.885	-.4034151	.3481837
wexp	-.1247765	.0203948	-6.12	0.000	-.1648403	-.0847127
ethblack	-.9180353	.2978136	-3.08	0.002	-1.503065	-.3330054
ethhisp	.4566623	.4464066	1.02	0.307	-.4202661	1.333591
sm	.3936096	.0378126	10.41	0.000	.3193298	.4678893
_cons	11.31124	.6172187	18.33	0.000	10.09876	12.52371

```
. predict Res1,residuals
```

Since the coefficient of Res_1 is not statistically significant, it would seem that schooling is not an endogenous variable. But we should not take these results at face value, because we have cross-sectional data and heteroscedasticity is usually a problem in such data. Therefore, we need to find a heteroscedasticity-corrected standard error, such as the HAC standard errors discussed in Chapter 5. The results are as shown in Table 6.7.

Now the coefficient of the Res_1 variable is statistically significant, at about the 8% level, indicating that education (schooling) seems to be endogenous. But note that we have dropped the variables *ethblack* and *ethhisp*, which might explain why the Res_1 variable is now significant (at the 8% level). As an exercise, see if the results change if the two dropped variables are included in the regression of Table 6.7.

Table 6.6 Hausman Test of Endogeneity of Schooling: Second-Step Results

```
regress lnearnings s female wexp ethblack ethhisp Res1

      Source |       SS           df       MS              Number of obs   =       540
-------------+----------------------------------           F(6, 533)       =     50.80
       Model |  77.8586986         6   12.9764498          Prob > F        =    0.0000
    Residual |  136.151689       533   .255444069          R-squared       =    0.3638
-------------+----------------------------------           Adj R-squared   =    0.3566
       Total |  214.010387       539   .397050811          Root MSE        =    .50541


-------------------------------------------------------------------------------
   lnearnings |     Coef.   Std. Err.      t     P>|t|     [95% Conf. Interval]
-------------+-----------------------------------------------------------------
           s |    .140068   .0224186     6.25    0.000     .0960283    .1841077
      female |  -.2997973   .044676     -6.71    0.000     -.38756    -.2120346
        wexp |   .0347099   .0056879     6.10    0.000     .0235365    .0458833
     ethblack |  -.1872501   .0752865    -2.49    0.013    -.3351448   -.0393554
     ethhisp |  -.0858509   .1013977    -0.85    0.398    -.2850391    .1133373
        Res1 |  -.0165025   .0245882    -0.67    0.502    -.0648041    .0317992
        _cons |   .4607717   .3765282     1.22    0.222    -.2788895    1.200433
-------------------------------------------------------------------------------
```

6.10.2 How to Find Whether an Instrument Is Weak or Strong

If an instrument used in the analysis is weak, in the sense that it is poorly correlated with the stochastic regressor for which it is an instrument, the IV estimator can be severely biased and its sampling distribution is not approximately normal, even in large samples. As a consequence, the IV standard errors and the confidence intervals based on them are highly misleading, leading to hypothesis tests that are unreliable.

To see why this is the case, refer to Equation (6.34). If ρ_{xz} in this equation is zero, the variance of the IV estimator is infinite. If ρ_{xz} is not exactly zero, but very low (the case of a weak instrument), the IV estimator is not

Table 6.7 Hausman Endogeneity Test With Robust Standard Errors

```
egress lnearnings s female wexp Res1,vce(robust)

Linear regression                          Number of obs   =       540

                                           F(4, 535)       =     59.14

                                           Prob > F        =    0.0000

                                           R-squared       =    0.3562

                                           Root MSE        =    .50747


-------------------------------------------------------------------------------
             |               Robust
lnearnings   |    Coef.    Std. Err.      t    P>|t|     [95% Conf. Interval]
-------------+-----------------------------------------------------------------
          s  |  .1642758   .0209439     7.84   0.000    .1231334     .2054183

     female  | -.3002845   .0443442    -6.77   0.000   -.3873947    -.2131744

       wexp  |  .0390386   .0053869     7.25   0.000    .0284565     .0496207

       Res1  | -.0407103   .022955     -1.77   0.077   -.0858034     .0043828

      _cons  |  .0311987   .3380748     0.09   0.927   -.6329182     .6953156
-------------------------------------------------------------------------------
```

normally distributed, even in large samples. But how do we decide in a given case whether an instrument is weak?

In the case of a single endogenous regressor, a rule of thumb says that an F statistic of less than 10 in the first step of the Hausman test suggests that the chosen instrument is weak. If it is greater than 10, it probably is not a weak instrument.

On that score, in our example Sm (mother's schooling) seems to be a strong instrument for S because the value of the F statistic in the first stage of the two-stage procedure is about 35, which exceeds the threshold value of 10. But this rule of thumb, like most rules of thumb, should not be used blindly. Actually, the situation is much better because of the relationship between the F and t statistics, namely, $F_{1,k} = t_k^2$, which in this case is $(10.41)^2 \approx 108$ (see the t value of Sm from the first part of Table 6.5). Note that this relationship between the F and the t statistic holds true only in the case of a single regressor.

6.10.3 The Case of Multiple Instruments

Since there are competing instruments, education may be correlated with more than one IV. To allow for this possibility, we can include more than one instrument in the IV regression. This is often done with the aid of 2SLS that we just discussed.

Step 1: We regress the suspected endogenous variable on all the instruments plus all exogenous regressors, and obtain its estimated value from this regression.

Step 2: We then run the earnings regression on the regressors included in the original model but replace the education variable by its value estimated from the Step 1 regression.

We can replace this two-step procedure by a single step by invoking *Stata's ivreg* command by including several instruments simultaneously, as the following example demonstrates.

For our earnings regression, in addition to mother's education (*Sm*), we can include father's schooling (*Sf*) and the number of siblings as instruments in the regression of earnings on education (*S*), gender (*female*=1), years of work experience (*wexp*), and ethnicity (dummies for black and Hispanic).

Step 1: Regress schooling (*S*) on all the original (nonstochastic) regressors and the instruments. From this regression, we obtain the estimated value of *S*, and call it *S*-hat.

Step 2: We now regress earnings on female, *wexp*, ethnic dummies, and *S*-hat, the latter estimated from Step 1 (see Table 6.8). But we can avoid the two-step procedure by the *ivreg* command from Stata, as shown in Table 6.8.

Compared with a single instrument in Table 6.7, when we introduced multiple instruments, the coefficient of *S* has gone down a bit, but it is still higher than the OLS regression. But notice again that the relative standard error of this coefficient is higher than its OLS counterpart, again reminding us that IV estimators may be less efficient.

6.10.4 Testing the Validity of Surplus Instruments

Earlier, we stated that the number of instruments must be at least equal to the number of stochastic regressors. So technically for our earnings regression, one instrument will suffice as in Table 6.7, where we used

Table 6.8 Earnings Function With Several Instruments

```
ivreg lnearnings female wexp ethblack ethhisp (s=sm sf siblings),robust
```

Instrumental variables (2SLS) regression	Number of obs	=	540
	F(5, 534)	=	26.63
	Prob > F	=	0.0000
	R-squared	=	0.3492
	Root MSE	=	.51071

```
--------------------------------------------------------------------------------
             |              Robust
lnearnings   |    Coef.    Std. Err.      t    P>|t|     [95% Conf. Interval]
-------------+------------------------------------------------------------------
          s  |   .1579691   .0216708    7.29   0.000     .1153986    .2005396
      female |  -.2976888   .0441663   -6.74   0.000    -.3844499   -.2109278
        wexp |   .0372111   .005846     6.37   0.000     .0257271    .0486951
     ethblack | -.1627797   .0625499   -2.60   0.010    -.2856538   -.0399056
     ethhisp |  -.0676639   .098886    -0.68   0.494    -.2619172    .1265893
        _cons |   .1689835   .3621567    0.47   0.641    -.542443     .8804101
--------------------------------------------------------------------------------
```

Instrumented: s

Instruments: female wexp ethblack ethhisp sm sf siblings

Sm (mother's education) as an instrument. In Table 6.8, we have three instruments, two more than the absolute minimum. How do we know that they are valid in the sense they are correlated with education but are not correlated with the error term? In simple terms, are they relevant?

As noted earlier, if the number of instruments is equal to the number of endogenous regressors, the regression coefficients are identified. But if the instruments are fewer than the number of endogenous variables, the regression coefficients are underidentified. On the other hand, if the number of instruments is greater than the number of endogenous regressors, the regression coefficients are overidentified.

In the present example, if we use three instruments (*Sm*, *Sf*, *siblings*), we have two extra or surplus instruments. How do we find out the validity of the extra instruments? We can proceed as follows:

1. Obtain the IV estimates of the earnings regression coefficients, including all the (exogenous) variables in the model plus all the instruments, three in the present case.

2. Obtain residuals from this regression; call them *Res*.

3. Regress *Res* on all the original regressors, including the instruments, and obtain the R^2 value from this regression.

4. Multiply the R^2 value obtained in Step 3 by the sample size ($n = 540$). That is, obtain nR^2. If all the surplus instruments are valid, it can be shown that $nR^2 \sim \chi^2_m$, that is, nR^2 follows the chi-square distribution with m *df*, where m is the number of surplus instruments, 2 in our case.

5. If the estimated chi-square value exceeds the critical chi-square value, say, at the 5% level, we conclude that at least one surplus instrument is not valid.

We have already given the IV estimates of the earnings regression, including the three instruments in Table 6.8. From this regression, we obtained the following regression as per Step 3 above. The results are given in Table 6.9.

We need not worry about the coefficients in this table. The important entity here is R^2, which is 0.0171. Multiplying this by the sample size of 540, we obtain $nR^2 = 9.234$. The chi-square 1% significance value for 2 *df* is about 9.21. So the computed chi-square value is highly significant, which suggests that at least one surplus instrument is not valid. We could throw away two of the three instruments, as we need just one to identify (i.e., estimate) the parameters. Of course, it is not a good idea to throw away instruments. There are procedures in the literature to use WLS to obtain consistent IV estimates. We leave it to the reader to discover more about this in the references.[18]

6.11 Regression Involving More Than One Endogenous Regressor

So far, we have concentrated on a single endogenous regressor. How do we deal with a situation of two or more stochastic regressors? Suppose in our

---

[18]For additional details, see Stock, J. H., & Watson, M. W. (2011). *Introduction to econometrics* (3rd ed., chap. 12). Boston, MA: Addison-Wesley.

Table 6.9 Test of Surplus Instruments

```
ivregress Res  female wexp ethblack ethhisp sm sf siblings
```

Source	SS	df	MS			
Model	2.38452516	7	.340646452			
Residual	136.894637	532	.257320746			
Total	139.279162	539	.258402898			

Number of obs =	540
F(7, 532) =	1.32
Prob > F =	0.2366
R-squared =	0.0171
Adj R-squared =	0.0042
Root MSE =	.50727

Res	Coef.	Std. Err.	Z	P>|Z|	[95% Conf.	Interval]
female	-.0067906	.0449329	-0.15	0.880	-.0950584	.0814771
wexp	-.0001472	.0047783	-0.03	0.975	-.0095339	.0092396
ethblack	-.0034204	.0708567	-0.05	0.962	-.1426136	.1357728
ethhisp	-.0197119	.1048323	-0.19	0.851	-.225648	.1862241
sm	-.0206955	.0110384	-1.87	0.061	-.0423797	.0009887
sf	.0215956	.0082347	2.62	0.009	.0054191	.0377721
siblings	.0178537	.0110478	1.62	0.107	-.0038489	.0395563
_cons	-.0636028	.1585944	-0.40	0.689	-.3751508	.2479452

earnings regression we think that the regressor work experience (*wexp*) is also stochastic. Now we have two stochastic regressors, education (*S*) and *wexp*. We can use the 2SLS method to handle this case.

Just as one instrument (*Sm*) sufficed to identify the impact of education on earnings, we need another instrument for *wexp*. We have a variable *age* in our data. So we can use it to proxy *wexp*. We can treat age as truly exogenous. To estimate the earning regression with two stochastic regressors, we proceed as follows:

Stage 1: We regress each endogenous regressor on all exogenous variables and the instruments and obtain the estimated values of these regressors.

Stage 2: We estimate the earnings function using all exogenous variables and the estimated values of the endogenous regressors from Stage 1.

However, we do not have to go through this two-stage procedure, for packages such as *Stata* can do this in one step. The results are given in Table 6.10. This regression shows that the return to education per incremental year is about 13.4%, ceteris paribus. The regressors, *female* and *ethblack*, are individually highly significant, as before, but the work experience variable is not statistically significant.

We have argued that IV estimation will give consistent estimates in case a regressor has serious measurement errors, even though the estimates thus obtained are inefficient. But if measurement errors are absent, OLS and IV

Table 6.10   IV Estimation With Two Endogenous Regressors

```
ivreg lnearnings female ethblack ethhisp (s wexp= sm age)

Instrumental variables (2SLS) regression
```

Source	SS	df	MS			
				Number of obs	=	540
				F(5, 534)	=	27.59
Model	73.628356	5	14.7256712	Prob > F	=	0.0000
Residual	140.382031	534	.262887699	R-squared	=	0.3440
				Adj R-squared	=	0.3379
Total	214.010387	539	.397050811	Root MSE	=	.51273

lnearnings	Coef.	Std. Err.	t	P>|t|	[95% Conf. Interval]	
s	.1338489	.0230934	5.80	0.000	.0884839	.1792139
wexp	.0151816	.0159219	0.95	0.341	-.0160957	.0464588
female	-.3378409	.053815	-6.28	0.000	-.443556	-.2321259
ethblack	-.215774	.079171	-2.73	0.007	-.3712987	-.0602493
ethhisp	-.1252153	.1069831	-1.17	0.242	-.3353745	.084944
_cons	.8959276	.4991938	1.79	0.073	-.084697	1.876552

```
Instrumented:  s wexp

Instruments:   female ethblack ethhisp sm age
```

estimates are both consistent, in which case we should choose OLS because it is more efficient. Thus, it behooves us to find out if the instruments chosen for consideration are valid.

A test developed by Durbin, Wu, and Hausman (DWH), but popularly known as the **Hausman test**, is one that is used in applied regression analysis to test the validity of instruments.

Although the mathematics of the test is involved, the basic idea behind the DWH test is quite simple. We compare the differences between OLS and IV coefficients of all the variables in the model, and obtain, say, $m = (b^{OLS} - b^{IV})$. Under the null hypothesis that $m = 0$, it can be shown that m is distributed as the chi-square distribution with degrees of freedom equal to the number of coefficients compared. If m turns out to be zero, it would suggest that the (stochastic) regressor is not correlated with the error term, and we can use OLS in lieu of IV, because OLS estimators are more efficient.

The results of the DWH test based on *Stata* are given in Table 6.11. In this table, the column (b) gives the estimates of the model under IV (*earniv*) and column (B) gives the estimates obtained by OLS (*earnols*). The next column gives the difference between the two sets of coefficients (m) and the last column gives the standard error of the difference between the two estimates.

We do not reject the null hypothesis that the OLS and IV estimates are statistically the same, for the probability of obtaining a chi-square value of

Table 6.11　The DWH Test of Instrument Validity for the Earnings Function

| | ── Coefficients ── | | | |
| | (b) | (B) | (b-B) | sqrt(diag(V_b-V_B)) |
	z3	z4	Difference	S.E.
s	.1579691	.1263493	.0316198	.0181687
female	-.2976888	-.3014132	.0037243	.0069645
wexp	.0372111	.0327931	.004418	.0026348
ethblack	-.1627797	-.2060033	.0432236	.0268566
ethhisp	-.0676639	-.0997888	.0321249	.0235867

$$\text{chi2}(5) = (b-B)'[(V\_b-V\_B)^{\wedge}(-1)](b-B)$$
$$= 3.03$$
$$\text{Prob>chi2} = 0.6955$$

Note: Z_3 = earniv; Z_4 = earnols.

b = consistent under H_0 and H_a obtained from *ivreg*.

B = inconsistent under H_a, efficient under H_0 obtained by OLS.

Test: H_0: difference in coefficients is not systematic.

3.03 or greater is about 69%. In this case, we should choose the OLS estimators, as they are more efficient than the IV estimators.

Although we have not considered all the data given in the NLSY79, based on the model considered here, it seems that the education variable (S) is probably not correlated with the error term. But the reader is advised to try other models from the data in NLSY79 to see if they arrive at a different conclusion.

It should be noted that the main purpose of the preceding exercise was to illustrate the mechanics underlying the various aspects of IV estimation. The results shown in the various tables are for illustrative purposes only, for we could choose different sets of instruments and possibly obtain different results than those presented here. Besides, we have only a comparatively small sample of data. With a much larger sample, one could introduce more instruments to find out the extent to which the results discussed here are substantiated or refuted.

6.12 Summary

One of the critical assumptions of the CLRM is that the error term and regressor(s) are uncorrelated. But if they are correlated, then we call such regressor(s) stochastic or endogenous regressors. In this situation, the OLS estimators are biased, and the bias does not disappear even if the sample size increases indefinitely. In other words, the OLS estimators are not even consistent. As a result, tests of significance and hypothesis testing become suspect.

If we can find proxy variables such that the instruments are uncorrelated with the error term, but are correlated with the stochastic regressors and are not candidates in their own right in the regression model, we can obtain consistent estimates of the coefficients of the suspected stochastic regressors. Such variables, if available, are called IVs, or instruments for short.

In large samples, IV estimators are normally distributed with mean equal to the true population value of the regressor under stress and the variance that involves the population correlation coefficient of the instrument with the suspect stochastic regressor. But in small, or finite, samples, IV estimators are biased and their variances are less efficient than the OLS estimators.

The success of IV depends on how strong the instruments are—that is, how strongly they are correlated with the stochastic regressor(s). If this correlation is high, we say such instruments are strong, but if it is low, we call them weak instruments. The efficiency of IV estimators depends directly on the degree of correlation between the instruments and the stochastic regressors. If the instruments are weak, IV estimators may not be normally distributed even in large samples. *Finding "good" instruments is not easy.* It requires intuition, introspection, familiarity with prior empirical work, or sometimes just luck. That is why it is important to test explicitly whether the chosen instrument is weak or strong, using tests such as the Hausman test.[19]

We need one instrument per stochastic regressor. But if we have more than one instrument for a stochastic regressor, we have a surfeit of instruments and we need to test their validity. Validity here means whether the surfeit instruments have high correlation with the regressor but are uncorrelated with the error term. Fortunately, several tests are available to test for this.

If there is more than one stochastic regressor in a model, we will have to find an instrument for each stochastic regressor. Again, we need to test the instruments for their validity.

One practical reason why IVs have become popular is that we have excellent statistical packages, such as *Stata* and *Eviews*, which make the task of estimating IV regression models very easy, provided that we have good instruments.

The topic of IV is still evolving, and considerable research is being done on it by various academics. It pays to visit their websites to learn more about the recent developments in the field. Of course, the Internet is an invaluable source of information on IV and other statistical techniques.

Appendix 6A: Properties of OLS When Random X and u Are Independently Distributed[20]

For the LRM $y = XB + u$, the usual OLS formulas are

$$b = (X'X)^{-1}X'y = Ay; A = (X'X)^{-1}X' \tag{6A.1}$$

$$e = My; M = [I - X(X'X)^{-1}X'] \tag{6A.2}$$

Estimate of σ^2:

$$S^2 = e'e / n - k \tag{6A.3}$$

$$\text{cov}(b) = \frac{S^2}{n}Q^{-1}; Q = (X'X/n) \tag{6A.4}$$

---

[19]For examples of successful IV regressions, see Angrist, J. D., & Pischke, J.-S. (2009). *Mostly harmless econometrics: An empiricist's companion* (chap. 4). Princeton, NJ: Princeton University Press.

[20]See Goldberger, A. S. (1991). *A course in econometrics* (chap. 25). Cambridge, MA: Harvard University Press.

But note that the matrices Q, A, and M are now *random* because they are functions of the random X, but conditional on X they are constant.

Under the classical LRM, we know that

$$E(b)=B; \text{var}(b)=\frac{\sigma^2}{n}Q^{-1}; E(e)=0; \text{var}(e)=\sigma^2 MM'=\sigma^2 M$$

M being idempotent. But when X is random, we have to note that

$$E(b|X)=E(Ay|X)=AE(y|X)=AXB=B \tag{6A.5}$$

$$\text{var}(b|X)=\text{var}(Ay|X)=A\text{var}(y|X)A'=\sigma^2 AA'=\frac{\sigma^2}{n}Q^{-1} \tag{6A.6}$$

$$E(e|X)=E(My|X)=ME(y|X)=MXB=0 \tag{6A.7}$$

$$\text{var}(e|X)=\text{var}(My|X)=M\text{var}(y|X)M'=\sigma^2 MM'=\sigma^2 M \tag{6A.8}$$

From the preceding it follows that

$$E(e'e|X)=\sigma^2\text{tr}(M)=\sigma^2(n-k) \tag{6A.9}$$

$$E(S^2|X)=\sigma^2 \tag{6A.10}$$

$$E[\text{var}(b)|X]=E(S^2 Q^{-1}|X)=E(S^2|X)Q^{-1}=\sigma^2 Q^{-1} \tag{6A.11}$$

In sum, conditional on any value of X, OLS estimators of the LRM parameters are unbiased. In essence, we are treating X as nonstochastic for all practical purposes. But we must state expressly that this result holds conditional on every value of X.

Furthermore, using the **law of iterated expectations**, it can be shown that the OLS parameters of the LRM are also unbiased unconditionally. Additionally, a version of the Gauss–Markov theorem states that in the class of estimators, conditional on every X that is linear and unbiased, the OLS estimator has minimum variance. As Goldberger states, "nothing in the randomness of the explanatory variables per se creates an objection to LS estimation."[21]

These preceding moments are conditional moments, conditional on the given values of X. But we can also obtain unconditional moments, using the law of iterated expectation, also known as the **law of total expectation**. To explain this law, consider a random variable Y and its unconditional expectation $E(Y)$ and also consider its conditional expectation based on

---

[21]Goldberger, A. S. (1991). *A course in econometrics* (p. 269). Cambridge, MA: Harvard University Press.

another random variable, $E(Y|X)$. The relationship between these two expectations is

$$E(Y)=E_X[E(Y|X)]$$

In other words, the marginal or unconditional expectation of Y is equal to the expectation of its conditional expectation, E_X, denoting that the expectation has taken over the values of X. Put simply, this law states that if we first obtain $E(Y|X)$ as a function of X and then take its expected or average value over the distribution of X values, we wind up with $E(Y)$.

Using this concept, we can show that[22]

$$E(b) = E_X[E(b|X)]=E_X(B)=B, \quad \text{using}\,(6A.5) \tag{6A.12}$$

$$\begin{aligned}\text{var}(b)&=E_X[\text{var}(b|X)] + \text{var}_X[E(b|X)]\\ &=E_X(\sigma^2 Q^{-1})+\mathbf{0}=\sigma^2 E(Q^{-1})\ (\text{see Equation 6A.6})\end{aligned} \tag{6A.13}$$

$$E(S^2)=E_X[E(S^2|X)]=E_X(\sigma^2)=\sigma^2 \tag{6A.14}$$

$$E[\text{var}(b)]=E_X\left\{E[\ \text{var}(b)|X]\right\} = E_X(\sigma^2 Q-1)=\sigma^2 E(Q^{-1}) \tag{6A.15}$$

What all this shows is that the OLS estimators are unbiased conditionally as well as unconditionally. Thus, nothing in the randomness of X prevents us from using OLS. The Gauss–Markov theorem also holds true in that conditional on every X in the class of estimators that are linear and unbiased, the OLS estimators have minimum variance.

Normality assumption

In the preceding derivations, we did not assume any probability distribution for the error term u. But if we assume that $u|X \sim N(0, \sigma^2)$, then $y|X \sim N(X, \sigma^2)$, so hypothesis-testing procedures follow the classical procedure. The main thing we have to emphasize is that we have to interject "$|X$" all throughout.

Appendix 6B: Properties of OLS Estimators When Random X and u Are Contemporaneously Uncorrelated

Here we have a situation in which the random X is uncorrelated with the error term at the same point in time. What this means is that

---

[22]Goldberger, A. S. (1991). *A course in econometrics* (p. 268). Cambridge, MA: Harvard University Press.

$$\text{cov}(X_1, u_1) = \text{cov}(X_2, u_2) = \ldots = \text{cov}(X_n, u_n) = 0$$

Consider the following simple example:

$$Y_t = B_1 + B_2 Y_{t-1} + u_t, \quad t = 1, 2, \ldots, n \tag{6B.1}$$

where Y_t = current consumption expenditure, Y_{t-1} = consumption expenditure in the previous period, and u_t is the error term.

Here Y_t is correlated with u_t, but as long as u_t is not autocorrelated, it is not correlated with the past value of Y_t, Y_{t-1}. The lagged Y_t is a predetermined value. This is what we mean by contemporaneously uncorrelated: u_t is correlated with Y_t but not with Y_{t-1}.

The OLS estimator of B_2 in (6B.1) is

$$b_2 = \frac{\sum_{t=2}^{T} y_t y_{t-1}}{\sum_{t=2}^{T} y_{t-1}^2} = B_2 + \frac{\sum_{t=2}^{T} y_{t-1} u_t}{\sum_{t=2}^{T} y_{t-1}^2} \tag{6B.2}$$

Now $E(b_2) \neq B$, because we cannot take the expectation of the ratio on the right-hand side of (6B.2), for E is a linear operator. Unless this last ratio term is zero, b_2 is a biased estimator of B_2. And there is no reason to believe that the last term is expected to be zero.

However, if the sample is large enough, we can use the $p$lim operation on (6B.2), which gives

$$p\lim(b_2) = B_2 + \frac{p\lim \Sigma y_{t-1} u_t}{p\lim \Sigma y_{t-1}^2}$$

$$= B_2 + \frac{\text{cov}(Y_{t-1}, u_t)}{\text{var}(Y_{t-1}^2)} = B_2 \tag{6B.3}$$

because the numerator of the ratio is zero.

In sum, although b_2 is biased in a finite sample, it is consistent in large samples under the assumed conditions. In other words, b_2 is a consistent estimator of B_2.

CHAPTER 7. SELECTED TOPICS
IN LINEAR REGRESSION

7.1 Introduction

In this chapter, we consider some of the frequently encountered topics in applied regression analysis. Specifically, we discuss the following topics:

1. Multicollinearity

2. Specification errors

3. Functional forms of LRMs

4. Qualitative regressors

5. Nonnormality of the error term

Although there are many other topics that could be discussed, I have confined myself to these topics not only because they have occurred frequently in applied research but also because their presence poses threats to statistical inference. Since the primary focus of this book is on LRMs—that is, models linear in their parameters—I have not discussed models that are nonlinear in their parameters because that would far expand the objectives of this book. Besides, a discussion of nonlinear-in-the-parameters regression models will require a book by itself.

7.2 The Nature of Multicollinearity[1]

Assumption 6 of the CLRM states that there is no **collinearity** or no **multicollinearity** if there is more than one exact relationship among the explanatory variables or the regressors. In matrix notation, this assumption means that the X matrix is of *full column rank*; that is, *the columns of the* X *matrix are linearly independent*. This requires that the number of observations, n, be greater than the number of parameters (i.e., the k regression coefficients) estimated. (See Appendix A on matrix algebra.)

There are two types of multicollinearity: (1) perfect and (2) imperfect. In perfect multicollinearity, there is one or more exact linear relationship among the regressors. This rarely happens in practice, but sometimes, it could happen

---

[1]For details and examples, see Gujarati, D. N., & Porter, D. (2009). *Basic econometrics* (5th ed., chap. 10, pp. 320–364). New York, NY: McGraw-Hill/Irwin.

by negligence in the structure of the X matrix. In case of imperfect multicollinearity, the regressors may be highly, but not perfectly, correlated. Let us consider the consequences of both perfect and imperfect multicollinearity.

7.2.1 Estimation in the Presence of Perfect Multicollinearity

Remember that the traditional OLS estimator is expressed as

$$b=(X'X)^{-1}X'y \tag{7.1}$$

and its variance is expressed as

$$\text{cov}(b)=\sigma^2(X'X)^{-1} \tag{7.2}$$

The CLRM assumes that the rank of the matrix X is k, the number of regressors in X. The rank of the $(X'X)$ matrix is also k. In this case, the $(X'X)$ matrix is called **nonsingular**. An implication is that the inverse of this matrix, $(X'X)^{-1}$, exists, which enables us to estimate B as shown in Equation (7.1). But in the case of perfect multicollinearity, the rank of the X is less than k, which means the matrix $(X'X)$ cannot be inverted, and therefore, we cannot estimate B. In this case, $(X'X)$ is called a **singular matrix**. (See Appendix A on matrix algebra.) In the case of perfect multicollinearity, it is futile to estimate B and the variances and standard errors of b, the estimator of B.[2]

Since the rank of X is less than k, the rank of $(X'X)$ is also less than k. Recall that $(X'X)b = X'y$, which is a system of k equations in k unknown (regression) coefficients. So we have a system of equations that has one or more redundant equations, which means there are more unknowns than the number of equations to estimate them. As a result, there is an infinite number of solutions for b. So the case of perfect multicollinearity is not of practical importance. But the case of imperfect multicollinearity often arises in practice. Therefore, we consider this case in some detail.

7.2.2 Estimation in the Presence of Imperfect or Near Multicollinearity

Since the $(X'X)$ matrix in this case is invertible, we can estimate b and its variance–covariance matrix from Equations (7.1) and (7.2). Not only is

---

[2]One way to find a solution to the problem is to use the **generalized inverse of a matrix**. This is a rather involved topic, and the interested reader may consult Schmidt, P. (1976). *Econometrics* (pp. 42–47). New York, NY: Marcel Dekker.

b still BLUE, but there are several consequences we have to consider, such as the following:

1. Although BLUE, the OLS estimators have large variances and covariances, making precise estimation difficult.

2. As a result, the confidence intervals tend to be much wider, leading to the nonrejection of the "zero null hypothesis" more readily.

3. As a result of the first consequence, the *t* ratios of one or more regression coefficients tend to be statistically insignificant. Even then R^2, the overall measure of goodness of fit, and the F statistic can be very high.

4. The OLS estimators and their standard errors can be sensitive to small changes in the data.

To see some of the consequences of collinearity, consider the following formula for the variance of b_j, the *j*th partial regression coefficient in the LRM:

$$\text{var}(b_j) = \frac{\sigma^2}{(1 - R_j^2)\Sigma x_j^2} \tag{7.3}$$

where R_j^2 is the coefficient of determination in the regression of the *j*th regressor on the other regressors in the model and $\Sigma x_j^2 = \sum_{i=1}^{n}(X_{ij} - \bar{X}_j)^2$. This equation shows that as $R_j^2 \to 1$, the variance of b_j increases, and in the limit when it is 1, its variance explodes toward infinity. Now consider the following expression:

$$\text{VIF}_j = \frac{1}{1 - R_j^2} \tag{7.4}$$

Therefore, we can write (7.3) as

$$\text{var}(b_j) = \frac{\sigma^2}{\Sigma x_j^2} \text{VIF}_j \tag{7.5}$$

VIF is called, appropriately, the **variance inflation factor** as it quantifies how rapidly the variance of b_j increases due to its linear dependence on the *j*th regressor with the other regressors in the model. A rule of thumb says that serious collinearity exists if the VIF >10.

Notice that the variance of b_j is proportional to σ^2 and the VIF but is inversely proportional to Σx_j^2. So whether var(b_j) is large or small will depend on three factors: (1) σ^2, (2) VIF, and (3) Σx_j^2. A high VIF can be countered by a low σ^2 or a high Σx_j^2 (i.e., a larger variability in the values of x_j) or both.

To shed more light on the collinearity problem, consider the following regression:

$$Y_i = B_1 + B_2 X_{2i} + B_3 X_{3i} + u_i \tag{7.6}$$

For this model, it can be shown that

$$\text{var}(b_2) = \frac{\sigma^2}{(1 - r_{23}^2)\Sigma x_{2i}^2} \tag{7.7}$$

$$\text{var}(b_3) = \frac{1}{(1 - r_{23}^2)\Sigma x_3^2} \tag{7.8}$$

where $x_{2i} = (X_{2i} - \bar{X}_2), x_{3i} = (X_{3i} - \bar{X}_3)$, and r_{23} = the correlation coefficient between X_2 and X_3.
   In this case, the VIF is

$$\text{VIF} = \frac{1}{(1 - r_{23}^2)}$$

Therefore, the variances given in Equations (7.7) and (7.8) can be written as

$$\text{var}(b_2) = \frac{\sigma^2}{\Sigma x_{2i}^2}\text{VIF} \tag{7.9}$$

$$\text{var}(b_3) = \frac{\sigma^2}{\Sigma x_{3i}^2}\text{VIF} \tag{7.10}$$

To see the speed with which the variances of the two coefficients increase, consider the VIF values of these coefficients for selected values of r_{23} given in the following table.

r_{23}	0	0.5	0.7	0.9	0.95	0.99	0.995
VIF	1	1.33	1.96	5.26	10.26	50.25	100

As this table shows, when $r_{23} = 0.5$, the VIF is 1.33, but when it is equal to 0.95, the VIF increases to 10.26, and when the correlation coefficient nears 0.995, the VIP jumps to 100.

If we want to establish, say, a 95% confidence interval for b_2 or b_3, for various values of r_{23}, we have the following outcome.

r_{23}	95% Confidence Interval for b_2
0	$b_2 \pm 1.96 se(b_2)$
0.5	$b_2 \pm 1.96\sqrt{1.33} se(b_2)$
0.95	$b_2 \pm 1.96\sqrt{10.2} se(b_2)$
0.995	$b_2 \pm 1.96\sqrt{100} se(b_2)$

Holding σ^2 and Σx_j^2 constant, it is clear from this table that the confidence interval widens rapidly as the VIF increases, which means increasingly one cannot reject the zero null hypothesis.

It may be noted that the inverse of the VIF is called **tolerance (TOL)**, that is,

$$TOL = \frac{1}{VIF} \qquad (7.11)$$

As you can see from Equation (7.4), when $R_j^2 = 1$, TOL = 0, and when it is 0, TOL = 1.

In the opinion of the late Professor Arthur Goldberg, the real problem with perfect or near-perfect multicollinearity is **micronumerosity** or the problem of "small sample size."[3] In regression analysis, it is assumed that n, the sample size, is greater than k, the number of regressors. Exact micronumerosity arises when $n = 0$. Near-micronumerosity arises when the rank condition $n > 0$ is barely satisfied. Put simply, high collinearity means that there is not enough information in the sample to estimate the parameters of the regression model. This does not mean that there is a literal connection between the sample size and collinearity, for collinearity can exist even in large samples.

---

[3]See Goldberger, A. S. (1991). *A course in econometrics* (pp. 248–250). Cambridge, MA: Harvard University Press.

7.2.3 Detection of Multicollinearity

There are several indicators of multicollinearity, and it is often not clear which is the best indicator. These indicators are as follows.

1. *High R^2 but few significant t ratios:* This is usually an indicator of collinearity.

2. *High pairwise correlation among the regressors:* A pairwise correlation between variables Y and X is measured as $r_{xy} = \dfrac{\Sigma x_i y_i}{\sqrt{\Sigma x_i^2 \Sigma y_i^2}}$, where the variables are measured as deviations from their mean value: $x_i = (X_i - \bar{X}), y_i = (Y_i - \bar{Y})$. But in computing such pairwise correlations, it is assumed that other variables in the model are held constant.

3. *Examination of partial correlation coefficients:* Unlike simple, or pairwise, correlation coefficients, partial correlation coefficients take into account the presence of other variables in the model. For example, in the regression (7.6), we have three pairwise correlation coefficients, r_{12}, r_{13}, and r_{23}, where the subscripts 1, 2, and 3, refer to the variables Y, X_2, and X_3. Suppose we want to find the correlation between variables 1 and 2, holding the influence of the variable 3 constant, which we denote as $r_{12.3}$, which we call a partial correlation coefficient. In computing this correlation, we hold the influence of the variable X_3 constant. The formula for this is $r_{12.3} = \dfrac{r_{12} - r_{13}r_{23}}{\sqrt{(1 - r_{13}^2)(1 - r_{23}^2)}}$, which expresses the partial correlation in terms of the simple, or pairwise, correlations. As you can see, $r_{12.3}$ is not the same as r_{12}, unless r_{13} and r_{23} are zero. So you can see that pairwise correlations as indicators of collinearity do not provide a good measure. Even if we compute partial correlations, there is no guarantee that they will provide an infallible guide to multicollinearity. Besides, if a regression model contains several regressors, computing several partial correlations may not be practical.

4. *Auxiliary regressions:* In the general k-variable LRM, there are k regressors. To see which of the regressors may be highly related to the other regressors in the model, we regress each regressor on the other regressors in the model and obtain the corresponding R^2 values. There will be $(k - 1)$ such regressions; they are called auxiliary because they are secondary to the main LRM. We can test the statistical significance of each of the R^2 values by the usual F test.[4] If one

---

[4]For details, see Gujarati, D. N., & Porter, D. (2009). *Basic econometrics* (5th ed., chap. 10, p. 339). New York, NY: McGraw-Hill/Irwin.

of the R^2 values is statistically significant, it would suggest which regressor is a linear combination of the other regressors in the model. In that case, we might think of dropping that variable from the model because it is represented by a (linear combination) of the other variables in the model. Instead of testing the statistical significance of all auxiliary R^2 values, one may adopt **Klein's rule of thumb**, which suggests that multicollinearity may be troublesome problem only if the R^2 obtained from an auxiliary regression is greater than the overall R^2, that is, that obtained from the regression of Y on all the regressors in the model.[5] Of course, like other rules of thumb, this one should be used judiciously.

5. *Tolerance and VIF:* Some authors use VIF as an indicator of multicollinearity. As a rule of thumb, if the VIF of a variable exceeds 10, which will happen if R_j^2 exceeds 0.90, that variable is said to be highly collinear.[6] Because of the intimate connection between tolerance and VIF, we can also use tolerance. But this indicator of collinearity is not free of criticism, for we have seen that the variance of an OLS estimator b_j depends on three factors: σ^2, Σx_j^2, and VIF. A high VIF can be counterbalanced by a low σ^2 or a high Σx_j^2. Therefore, a high VIF is neither necessary nor sufficient to get high variances and high standard errors. In all this discussion, the terms *high* or *low* are used in a relative sense.

7.2.4 Remedial Measures

The remedies suggested in the literature are essentially ad hoc. Some of these are as follows.

1. *A priori information about one or more regression coefficients:* Suppose in regression (7.6) prior empirical work shows that $b_3 = 0.6$. In this case, we can subtract $0.6X_3$ from Y_i. So there is no collinearity problem in this case. Of course, in a regression involving several regressors, this remedy may not always be practical.

2. *Dropping one or more collinear variables:* If auxiliary regressions suggest that one or more regressors are highly collinear with the other regressors in the model, one may think of dropping those variables from the model. But one has to guard against model

[5]Klein, L. (1962). *An introduction to econometrics* (p. 101). Englewood Cliffs, NJ: Prentice Hall.

[6]Kleinbaum, D., Kupper, L. L., & Muller, K. E. (1988). *Applied regression analysis and other multivariate methods* (2nd ed., p. 210). Boston, MA: PWS-Kent.

specification errors, a topic discussed in the next section. While multicollinearity may prevent precise estimation of the parameters of the regression model, omitting one or more variables may seriously mislead us as to the true values of the parameters. But recall that OLS estimators are BLUE despite near collinearity.

3. *Transformation of variables:* Suppose in the regression model (7.6) Y represents consumption expenditure in dollars, X_2 income in dollars, and X_3 the total population. Since income and population grow over time, they are likely to be correlated. One "solution" to this situation is to express both consumption expenditure and income on a per capita basis, that is, dividing the equation on both sides by the total population to obtain

$$\frac{Y_i}{X_{3i}} = B_1 \left(\frac{1}{X_{3i}} \right) + B_2 \left(\frac{X_{2i}}{X_{3i}} \right) + B_3 + \left(\frac{u_i}{X_{3i}} \right) \tag{7.12}$$

Such a transformation may reduce collinearity in the original variables. Note that in this transformation the "intercept" B_3 is in fact the coefficient of the inversed population variable. In effecting such transformations, we have to guard against creating other problems. If the original error term u_i is homoscedastic, the transformed error term (u_i / X_{3i}) will be heteroscedastic.

4. *Acquiring additional data:* Recall the discussion on micronumerosity. It is possible the problem of multicollinearity is due to the smallness of the sample size. To see this, in the three-variable regression (7.6) we showed that

$$\text{var}(b_2) = \frac{\sigma^2}{(1 - r_{23}^2) \sum x_{2i}^2}$$

Now as the sample size increases, Σx_{2i}^2 will generally increase, therefore for any given r_{23}, the variance of b_2 will decrease, thus decreasing the standard error that will enable us to estimate B_2 more precisely.

As an illustration, consider the following regression of consumption expenditure Y on income X_2 and wealth X_3 based on 10 observations.[7] A priori, income and wealth both are expected to have a positive impact on Y.

$$\hat{Y}_i = 24.377 + 0.8716 X_{2i} - 0.0349 X_{3i}$$
$$t = (3.875) \quad (2.7726) \quad (-1.1595) \quad R^2 = 0.9682 \tag{7.13}$$

[7] I am grateful to my former colleague the late Albert Zucker for providing me with the results given in the following two regressions.

where t is the t statistic obtained under the null hypothesis that the relevant population coefficient is zero (the zero null hypothesis). The hat over Y represents the estimated Y values from the regression.

In this regression, contrary to prior expectations, the wealth coefficient not only is statistically insignificant but also has the wrong sign. But when the sample size was increased to 40 observations, the following results were obtained:

$$\hat{Y}_i = 2.0907 + 0.7299X_{2i} + 0.0605X_{3i}$$
$$t = (0.8713)\,(6.0014) \quad (2.0014) \quad R^2 = 0.9672$$

(7.14)

Now both coefficients are positive and statistically significant.

Obtaining additional or "better" data is not always easy because of the costs involved and because most applied researchers depend on publicly available data.

Ridge regression: Originally suggested by Hoerl and Kennard, the $(X'X)$ matrix is manipulated to obtain reduced variances of the OLS estimators.[8] The technical aspects of this method are discussed in Appendix 7A.

Principal components analysis (PCA): This is a statistical technique that can transform correlated variables into *orthogonal or uncorrelated variables*. The variables thus transformed are called **principal components**.[9] Regressions can then be run on the principal components. The practical problem is how to interpret the principal components. That is why this method is not much used in economics.

7.3 Model Specification Errors

Assumption 7 of the LRM assumes that the regression model used in the analysis is "correctly specified." By correct specification, we mean one or more of the following:

1. The model does not exclude any "core" variables.

2. The model does not include superfluous variables.

3. The functional form of the model is suitably chosen.

---

[8]Hoerl, A. E., & Kennard, R. W. (1970). Ridge regression: Biased estimation for nonorthogonal problems. *Technometrics, 12,* 55–67.

[9]For details of this technique, see Gujarati, D. (2015). *Econometrics by example* (2nd ed., pp. 89–92). London, England: Palgrave Macmillan.

4. There are no errors of measurement in the regressand and regressors.

5. Outliers in the data, if any, are taken into account.

6. The probability distribution of the error term is well specified.

A detailed discussion of these types of errors will take us far afield.[10] To give the flavor of the problems involved, we discuss three cases: (1) exclusion of the "core" or relevant regressors, (2) inclusion of the superfluous variables, and (3) the functional form of regression models.

7.3.1 Exclusion of Relevant Variables: Underfitting a Model

We start with the CLRM

$$y = XB + u \tag{7.15}$$

where X is of order $(n \times k)$ and B is of order $(k \times 1)$.
Let us partition this model as

$$y = (X_1, X_2) \begin{pmatrix} B_1 \\ B_2 \end{pmatrix}$$

$$y = X_1 B_1 + X_2 B_2 + u \tag{7.16}$$

where X_1 is of order $(n \times k_1)$, X_2 is of order $(n \times k_2)$, B_1 is of order $(k_1 \times 1)$, and B_2 is of order $(k_2 \times 1)$; $(k_1 + k_2) = k$.

7.3.1.1 Consequences of Underfitting a Model

The technical discussion of this specification error is given in Appendix 7B. In summary, the consequences of this specification error are as follows:

1. If the left-out, or omitted, variables are correlated with the variables included in the model, the coefficients of the estimated model are biased as well as inconsistent.

2. Even if the excluded variables are not correlated with the variables included in the model, the intercept of the estimated model is biased.

3. The error variance σ^2 is incorrectly estimated.

---

[10]For details and examples, see Gujarati, D. (2015). *Econometrics by example* (2nd ed., chap. 7). London, England: Palgrave Macmillan.

4. The variances of the estimated coefficients of the misspecified model are biased, which means the estimated standard errors are also biased.

5. As a result, the usual confidence intervals and hypothesis-testing procedures are unreliable, leading to misleading conclusions about the statistical significance of the estimated parameters.

6. In consequence, forecasts based on the incorrect model and the forecast confidence intervals are suspect.

Some of these consequences can be illustrated with the wage regression (4.3) that we discussed in Chapter 4. This regression expresses hourly wages in dollars as a function of gender, race, and union membership; education in years; and work experience in years. The results of this regression are given in Table 4.1.

The results show that each of the explanatory variables has a statistically significant impact on hourly wages.

To illustrate the impact of excluding relevant variables from the model, and to illustrate the method of restricted least squares, in Table 4.3 we presented the results of the wage regression without the inclusion of the two qualitative, or dummy, variables, race and union status—that is, the results of the restricted regression. The results of the two regressions show some subtle differences in the estimated coefficients, their standard errors, and the standard error of the regression. On the basis of the F test, we showed that we should not have excluded the two qualitative regressors from the wage regression.

7.3.2 Inclusion of Irrelevant Variables: Overfitting a Model

The technical details of this case are also shown in Appendix 7B, but we summarize the results as follows.

Knowing the consequences of excluding relevant variables from a model, one can go to extremes and include all sorts of variables in the belief that the irrelevant variables will not do much harm. In a sense, this is true, for it can be shown that the inclusion of irrelevant variables does not destroy the properties of unbiasedness and consistency of OLS estimators. The estimated error variance is also unbiased, which means that the standard hypothesis-testing procedures are valid. But inclusion of irrelevant variables involves some loss in the efficiency of the estimated parameters because we lose several degrees of freedom in estimating the coefficients of the "nuisance" variables. That is why the variances of the estimated coefficients are larger than those of the true model.

We can summarize the consequences of the two types of specification errors as shown in Table 7.1.

Table 7.1 Consequence of Specification Errors[11]

	$y = XB + u$	
State of models	True model: $y = X_1B_1 + X_2B_2 + u$	True model: $y = X_1B_1 + u$
Model chosen: $y = X_1B_1 + X_2B_2 + u$	b is minimum variance unbiased	b_1^* is unbiased but not minimum variance
Model chosen: $X_1B_1 + u$	b_1^* is biased, but its covariance is less than that of the unbiased estimator, but σ^{2*} is biased upward	b_1 is minimum variance unbiased

Note: b^* is the estimator of b under the wrong choice.

There are several diagnostic tests to detect one or more of the specification errors.[12]

7.3.3 Functional Forms of Regression Models[13]

The primary emphasis of this text is on LRMs, that is, regression models that are linear in their parameters; the dependent and explanatory variables need not be linear. We now discuss models that may be nonlinear in their variables but are linear in their parameters or that can be made so by suitable transformations of the variables. To ease the algebra, we consider bivariate regression models, although the discussion below can be extended to multiple regression models.

1. *Log-linear models:* Consider the following model, known as the **exponential regression model**:

$$Y_i = B_1 X_i^{B_2} e^{u_i} \tag{7.17}$$

---

[11]The following table is adapted from Judge, G. G., Griffiths, W., Carter Hill, R., & Lee, T.-C. (1980). *Theory and practice of econometrics* (p. 413). New York, NY: Wiley.

[12]See Gujarati, D. N., & Porter, D. (2009). *Basic econometrics* (5th ed., chap. 13). New York, NY: McGraw-Hill/Irwin.

[13]For details and a numerical example, see Gujarati, D. N., & Porter, D. (2009). *Basic econometrics* (5th ed., chap. 6). New York, NY: McGraw-Hill/Irwin.

which may be expressed as

$$\ln Y_i = \ln B_1 + B_2 \ln X_i + u_i \tag{7.18}$$

where ln = natural log, that is, log to the base e, where $e = 2.718^{14}$ which we can write as

$$\ln Y_i = A + B_2 \ln X_i + u_i \tag{7.19}$$

where $A = \ln B_1$.

This model is linear in A and B_2, is linear in the logarithms of Y and X_2, and can be estimated by OLS. Such a model is called a **log-log, double-log**, or **log-linear** model.

An important property of the log-linear model is that the slope coefficient B_2 measures the **elasticity of Y with respect to X**, that is, the percentage change in Y for a given (small) percentage change in X.[15]

Thus, if Y represents the quantity of Y demanded and X its unit price, B here represents the **price elasticity of demand**, a parameter of considerable economic interest. Model (7.18) is also known as the **constant elasticity model** because the elasticity remains constant no matter at which value of X we measure it.

Model (7.19) can be extended to have more than one regressor in the log form. As an example, consider the following model:

$$\ln Y_i = B_1 + B_2 \ln X_{2i} + B_3 \ln X_{3i} + u_i \tag{7.20}[16]$$

This model is an example of a logarithmic production function, where Y = output, X_2 = labor input, and X_3 = capital input. In this model, each slope coefficient measures the elasticity of output with respect to the relevant input.

As a concrete example, we estimated (7.20) for the U.S. economy based on the 50 U.S. states and Washington, D.C., for the year 2005. The output

---

[14]In practice, we may use common logarithms, that is, log to the base 10. The two logarithms are related in that $\ln_e X = 2.3026 \log_{10} X$. By convention, ln means natural log and there is no need to write the subscript e explicitly.

$$15 \quad \frac{d \ln Y}{d \ln X} = \frac{dY / Y}{dX / X} = \frac{dY}{dX} \frac{X}{Y}$$

[16]This is the logarithmic transformation of the well-known Cobb–Douglas function: $Y_i = A X_{2i}^{B_2} X_{3i}^{B_3} e^{u_i}$.

is measured by value added in thousands of dollars, labor input is measured in thousands of worker hours, and capital input is measured by capital expenditure in thousands of dollars. We are using each state as the observational unit. The number of observations here is 51. The fitted production function is as follows:

$$\ln Y_i = 3.8876 + 0.4683 \ln X_{2i} + 0.5212 \ln X_{3i}$$
$$t = (9.8115)\,(4.7341) \qquad (5.3802) \quad R^2 = 0.9641$$

(7.21)

Note: The t values are the t statistics, which are obtained by dividing the estimated coefficients by their standard errors under the null hypothesis that each of the true coefficients (i.e., the population coefficients) is zero—the zero null hypothesis. The estimated t values are highly significant, for their individual p values are extremely low.

The slope coefficients are elasticities. Thus, holding capital input constant, if the labor input goes up by, say, 1%, the average output goes up by about 0.46%. Similarly, holding the labor input constant, if the capital input increases by, say, 1%, the average output goes up by about 0.52%.

2. *Semilog models:* There are two types of semilog models: (1) log-lin and (2) lin-log. In log-lin models, the dependent variable is in the log form and the explanatory variables are in the linear or log form. In the lin-log model, the dependent variable is in the linear form and the explanatory variables are in the linear or log form.

a. *Log-lin model:* Consider the following model:

$$\ln Y_t = B_1 + B_2 X_t + u_t$$

(7.22)

In this model, the slope coefficient B_2 measures the *proportional* or *relative change in Y for a given absolute change in the value of the regressor,*[17] that is,

$$B_2 = \frac{\text{Relative change in } Y}{\text{Absolute change in } X}$$

(7.23)

If we multiply (7.23) by 100, we obtain the percentage change, or the **growth rate**, in Y for an absolute change in X. Model (7.22) is often called the **growth model**. If, say, Y is population in (7.22) and X is time measured chronologically,

---

[17] $\dfrac{d \ln Y}{dX} = \left(\dfrac{1}{Y}\right)\dfrac{dY}{dX} = \dfrac{(dY/Y)}{dX}$

we obtain the rate of growth in population. If you multiply B_2 by 100, then we obtain what is called the **semielasticity** of Y with respect to X.

As an example, consider the following model:

$$\ln Y_t = B_1 + B_2 X_t + u_t \tag{7.24}$$

where Y, say, is the real gross domestic product (RGDP), that is, GDP adjusted for inflation, and X is, say, time measured chronologically. Based on the RGDP for the United States for 1960 through 2007, we obtained the following regression results:

$$\ln \text{RGDP}_t = 7.8756 + 0.0314 \text{time}_t$$
$$t = (807.0072) \ (90.8165) \ R^2 = 0.9944 \tag{7.25}$$

This regression shows that for the period 1960 through 2007 RGDP of the United States had been increasing at the rate of about 3.14% and this growth rate is statistically significant as the t value is so large.

b. *Lin-log model:* Now consider the following model:

$$Y_i = B_1 + B_2 \ln X_i + u_i \tag{7.26}$$

In this model,

$$B_2 = \frac{\text{Change in } Y}{\text{Change in } \ln X} = \frac{\text{Change in } Y}{\text{Relative change in } X} \tag{7.27}^{[18]}$$

Therefore,

$$\Delta Y = B_2 (\Delta X / X) \tag{7.28}$$

where Δ denotes absolute change.

Note that a change in the log of a number is a relative change.

This equation states that the absolute change in Y is equal to slope times the relative change in X. Thus, if $(\Delta X/X)$ changes by 0.01 units (or 1%), the

---

[18] $\dfrac{dY}{dX} = B_2 \left(\dfrac{1}{X}\right)$. Therefore, $B_2 = \dfrac{dY}{\dfrac{dX}{X}}$. Note that $\dfrac{dX}{X}$ is a relative change.

absolute change in Y is $0.01(B_2)$. If in an application one finds $B_2 = 150$, the absolute change in Y is $(0.01)(150) = 1.5$.

The lin-log model has been used in **Engel expenditure functions,** named after the German statistician Ernst Engel (1821–1896), who found that the share of expenditure on food decreases as total expenditure increases. That is, as people become more affluent, comparatively speaking, they spend less on food but more on other items. As an illustration of this, see Exercise 7.1.

3. *The reciprocal model:* Consider the following model:

$$Y_i = B_1 + B_2 \left(\frac{1}{X_i} \right) + u_i \tag{7.29}$$

This is called a reciprocal model because the dependent variable is inversely related to the regressor X. Although nonlinear in variable X, this is an LRM because it is linear in the parameters B_1 and B_2. A feature of this model is that as X increases indefinitely, the term $B_2(1/X)$ approaches zero (*note*: B_2 is constant) and Y approaches the limiting or **asymptotic** value B_1. So models like (7.29) have built in them an asymptote or limit value that the dependent variable will take as the value of the X variable increases indefinitely.[19] As an example of this, see Exercise 7.2.

4. *Polynomial regression models:* Consider the following model:

$$Y_i = B_1 + B_2 X_i + B_3 X_i^2 + B_4 X_i^3 + u_i \tag{7.30}$$

This is called a polynomial regression model, here a third-degree polynomial. Such models have been used in economics to deal with cost and production functions. The model (7.30) is still an LRM because the parameters enter linearly.

To illustrate (7.30), let us revisit the constant rate of growth model of RGDP discussed in (7.25) and consider the following model of RGDP:

$$Y_t = B_1 + B_2 X_t + B_3 X_t^2 + u_t \tag{7.31}$$

---

[19]Note that for this model the slope is $\dfrac{dY}{dX} = -B_2 \left(\dfrac{1}{X^2} \right)$. So if B_2 is positive, the slope is negative throughout, and if B_2 is negative, the slope is positive throughout.

where Y_t is RGDP at time t and X_t and X_t^2 are time and time squared, respectively, time being measured chronologically. In our example, the time period is from 1960 to 2007.

Equation (7.31) is a quadratic, or second-degree polynomial, in time. We can call it the quadratic trend model.

The results of this model are as follows:

$$\text{RGDP}_t = 2651.3810 + 68.5343\text{time} + 2.4175\text{time}^2$$
$$t = (38.1543) \quad (10.4758) \quad (18.6764) \quad R^2 = 0.9967 \tag{7.32}$$

How do we interpret this regression? Let us take the derivative of (7.32) with respect to time, which is

$$\frac{d\text{RGDP}}{d\text{time}} = B_2 + 2B_3\text{time} \tag{7.33}$$

Using the coefficient values from regression (7.32), we get

$$\frac{d\text{RGDP}}{d\text{time}} = 68.5343 + 2(2.4175)\text{time}$$
$$= 68.5343 + 4.8350\text{time} \tag{7.34}$$

As (7.34) shows, the rate of change of RGDP depends on the time at which the rate of change is measured.

As the results of the constant growth rate model of (7.25) and the variable rate of change model of (7.34) show, there is more than one way of modeling a particular phenomenon.[20]

To sum up our discussion of the functional forms of regression models, there are a variety of LRMs that can be used in empirical work as long as they are linear in their parameters or can be so made with suitable transformations.

7.4 Qualitative or Dummy Regressors[21]

We continue with the standard LRM: $y = XB + u$. Qualitative, or dummy, variables taking values of 1 and 0 show how qualitative regressors can be

---

[20]For further discussion of the functional forms of the regression model, see Gujarati, D. (2015). *Econometrics by example* (2nd ed., chap. 2). London, England: Palgrave Macmillan.

[21]For details and examples, see Gujarati, D. (2015). *Econometrics by example* (2nd ed., chap. 3). London, England: Palgrave Macmillan.

"quantified" and the role they play in regression analysis. If there are differences in the response of the regressand because of qualitative regressors, they will be reflected in the differences in the intercepts, or the slope coefficients, or both of the various subgroup regressions.

We have already seen how these dummy variables work in the wage regression that we discussed in Chapter 4. There we saw that female wage earners earn about \$3 less per hour than their male counterparts, or that nonwhite workers on average earn \$1.50 less per hour than the white workers, holding all other variables constant. We can also allow for **interaction** between two qualitative variables or between a quantitative variable and a qualitative variable. For example, in the wage regression (4.3), we can add a variable (**FE × NW**), the interaction between female and nonwhite, the idea being to find out whether the hourly wage of a nonwhite female worker differs from the hourly wage of a female worker alone or being a nonwhite alone. Similarly, we can allow for interaction between education and gender. This will tell us if an educated female worker earns differently than an educated male worker.

Dummy variables have been used in a variety of applications, such as (1) comparing two or more regressions, for example, wage regressions in two or more industries; (2) structural breaks in data, because of, say, wars; (3) the analysis of seasonal data; and (4) piecewise linear regressions.

Dummy variables pose no special estimation problems, but care needs to be exercised in their use. First, if there is an intercept in the model, the number of dummy variables must be one less than the number of classifications of each qualitative variable. Second, the coefficient of the dummy variable must always be interpreted in relation to the *reference category*, that is, the category that receives the value of zero. Third, as noted earlier, dummy variables can interact with quantitative as well as with qualitative variables. Fourth, if a model has several qualitative variables with several categories, introduction of dummies for all the combinations can consume a large number of degrees of freedom, especially if the sample size is relatively small.

One can also use dummy variables for the regressand, but that takes us into nonlinear regression models, which is not the focus of this book.[22]

7.5 Nonnormal Error Term[23]

Recall that OLS does not require that the regression error term, u, be normally distributed. The OLS estimators under the standard assumptions are

---

[22]For this, the reader can refer to Gujarati, D. (2015). *Econometrics by example* (2nd ed., chaps. 8–12). London, England: Palgrave Macmillan.

[23]For proofs of the following statements, see Schmidt, P. (1976). *Econometrics* (pp. 55–64). New York, NY: Marcel Dekker.

BLUE. We explained the reason behind the use of the normality assumption and also stated that the commonly used significance tests, such as the t, F, and χ^2, are based on the normality assumption. With this assumption, we showed that the method of ML can be used as an alternative to OLS. We also showed that the ML estimators have several desirable statistical properties, especially in large samples.

Can we drop the normality assumption, and with what effect?

As long as we assume that the errors u_i are iid with zero mean and finite variance σ^2, we need not invoke the normality assumption, for the properties of the OLS estimators that they are unbiased, consistent, and BLUE with covariance matrix $\sigma^2(X'X)^{-1}$ and that S^2, the estimator of σ^2, is unbiased and consistent still hold without the normality assumption.

However, to make probabilistic statements about the OLS estimators, we need to find their (probability) distributions. Since these estimators depend on the distribution of the error term u, we need to find out its probability distribution. For reasons already discussed, it is usually assumed that $u \sim N(0,\sigma^2)$. The law of large numbers and the CLT can be used to justify this assumption.

But if u is not normally distributed, b is not normally distributed, $(n-k)\dfrac{S^2}{\sigma^2}$ does not have the chi-square distribution, and b and S^2 are not efficient in small as well as large samples. As a result, the usual tests of hypotheses and confidence intervals are not valid.

The situation is not completely hopeless. If we assume that the error terms, u_i, are iid with zero mean and finite variance σ^2 and if $Q = \lim(X'X/n)$ is finite and nonsingular, then

$$\frac{1}{\sqrt{n}} X'u \text{ converges in distribution to } N(0,\sigma^2 Q).^{24} \qquad (7.35)$$

Under the same conditions, it can be shown that

$$\sqrt{n}(b - B) \to N(0,\sigma^2 Q^{-1}) \qquad (7.36)$$

This shows that b has the same asymptotic distribution whether or not the error terms are normally distributed, provided they are iid with zero mean and finite variance. As a consequence, the usual t tests are *asymptotically* valid. Of course, we replace the t distribution with the standard normal distribution, $N(0, 1)$. This is also true of the usual F tests. It is important to note that one should not use these tests if the sample size is relatively small.

---

[24]For proof, see, Schmidt, P. (1976). *Econometrics* (pp. 56–60). New York, NY: Marcel Dekker.

7.5.1 Normality Test

The conventional tests of hypotheses and confidence interval estimates of the regression parameters are based on the assumption that the regression error term u is normally distributed. In practice, how do we find out whether this assumption is tenable? In the literature, there are several tests of normality, informal and formal.[25] Here, we will discuss a popularly used *large-sample* test of normality, the **Jarque–Bera (JB) test**.[26] The JB statistic is defined as

$$JB = n \left[\frac{S^2}{6} + \frac{(K-3)^2}{24} \right] \sim \chi_2^2 \qquad (7.37)$$

where n is the sample size, S is the skewness coefficient, and K is the kurtosis coefficient. As (7.37) shows, the JB statistic follows the chi-square distribution with 2 *df*. For a normally distributed variable, $S = 0$ and $K = 3$. Therefore, the closer the JB statistic is to zero, the better is the normality assumption. Of course, we can use the chi-square distribution to find the p value of the JB statistic.

Note that the JB statistic is a test of the joint hypothesis that $S = 0$ and $K = 3$. That is why the JB statistic has 2 *df*: because we impose two restrictions, that $S = 0$ and $K = 3$ simultaneously.

Remember that the normality assumption pertains to the population error term u, but in practice, all we have are the residuals e. Therefore, in practice, the JB statistic is based on the estimated u ($= e$), especially if the sample size is reasonably large. In an application, if we find that the computed JB statistic exceeds the critical chi-square value, say, at the 5% level, we reject the hypothesis that the regression error term is normally distributed.

7.6 Summary

In Chapter 2, we introduced the CLRM, which is the cornerstone of most linear regression theory. The CLRM is based on several assumptions. A practical question is "How realistic are these assumptions?" But note that in any scientific study, we make certain assumptions because they facilitate the development of the subject matter in gradual steps, not because they are necessarily realistic in the sense that they replicate reality. As one author

---

[25]See Ghasemi, A., & Zahedias, S. (2012). Normality tests for statistical analysis: A guide for non-statisticians. *International Journal of Endocrinology and Metabolism, 10*(2), 486–489.

[26]Jarque, C. M., & Bera, A. K. (1987). A test of normality of observations and regression residuals. *International Statistical Review, 55*, 163–172.

notes, "If simplicity is a desirable criterion of good theory, all good theories idealize and oversimply outrageously."[27]

In Chapters 2, 3, and 4, we presented the core linear regression theory based on the simplifying assumptions, and in Chapters 5 and 6, we critically examined some of the assumptions underlying the CLRM, namely, the assumption of the scalar covariance matrix of the regression error term and the assumption that the regressors and the regression error term are independent, or at least not correlated. We suggested some of the ways that the CLRM can be modified.

In this chapter, we examined some of the problems researchers face in applying the theory in practice. The topics we discussed in this chapter relate to multicollinearity, specification errors, functional forms of LRMs, qualitative regressors, and the nonnormality of the regression error term.

We showed that as long as multicollinearity is not perfect, the method of OLS can be applied in practice and the estimators thus obtained produce estimators that are BLUE, assuming all the other assumptions of the CLRM hold.

Model specification is a serious problem in practice if we omit the core variables from the model in that the OLS estimators and their variances are biased. On the other hand, if we include redundant regressors in the model, the OLS estimators still retain their BLUE property, ceteris paribus. The only penalty we pay is in the reduced precision of the estimated parameters.

In the confines of the linear-in-the-parameter regressions models, there are a variety of functional forms we can use in practice. In this chapter, we illustrated their specific features.

The regressors included in the LRM need not all be quantitative. Some of them can be qualitative, or dummy, variables. In this chapter, we showed how in practice such variables are used and how they are interpreted.

The traditional Z, t, F, and chi-square tests used in testing hypotheses in the CLRM are based on the assumption that the regression error term is normally distributed. In this chapter, we briefly discussed the consequences of violating the normality assumption and showed that in sufficiently large samples we can still use the classical tests of significance.

Exercises

7.1 Using the lin-log model of (7.26), we obtained the following regression results:

$$Y_i = 0.9304 - 0.0777 \ln X_i$$
$$t = (25.5836)(-21.6482) \quad R^2 = 0.3508$$

---

[27]Blaug, M. (1992). *The methodology of economics: Or how economists explain* (2nd ed., p. 92). New York, NY: Cambridge University Press.

where Y = share of food expenditure in total expenditure and X = total expenditure. The results are based on the data for 869 U.S. households in 1995.

a. How would you interpret this regression?

b. What is the interpretation of the slope coefficient of about −0.08?

c. How would you compute the elasticity of the share of food expenditure in relation to the total expenditure? (*Hint :* Elasticity $= \dfrac{dY}{dX}\dfrac{X}{Y}$)

7.2 Using the same data as in Exercise 7.1, we estimated the reciprocal model as in Equation (7.29) and obtained the following results:

$$Y_i = 0.0772 + 1331.3380 \left(\frac{1}{X_i} \right)$$

$$t = (19.2595)\ (20.8161)\ \ R^2 = 0.3332$$

a. How would you interpret this regression?

b. What is the interpretation of the slope coefficient?

c. Is the rate of change in the food expenditure in relation to total expenditure positive or negative throughout?

d. What would be the share of food expenditure in the total expenditure if the total expenditure were to increase indefinitely? (*Hint:* the intercept)

7.3 Suppose instead of estimating the quadratic trend model (7.31), we estimate the model without the quadratic term. The results are as follows:

$$RGDP_t = 1664.2180 + 186.9939 time_t$$

$$t = (12.6078) + (39.8717)\ \ R^2 = 0.9718$$

This model is known as the linear trend model.

a. How would you interpret this regression?

b. Between the quadratic trend model of Equation (7.32) and the linear trend model, which would you choose? And why?

c. If the quadratic trend model is the "true" model, what type of specification error is involved if you use the linear trend model?

7.4 In the wage regression results given in Table 4.1, the dependent variable was w, hourly wage rate in dollars. Suppose now we regress the log of the wage rate on the regressors given in Table 4.1. The results of this regression are given in the following table (*Note:* LW = natural log of W). In this regression, as in the results given in Table 4.1, FE (female), NW (nonwhite), and UN (union) are qualitative or dummy variables; ED (education in years) and EX (experience in years) are quantitative variables. The results of "log wage regression" are as follows:

Dependent Variable: LW

Method: Least Squares

Date: 02/25/17 Time: 17:55

Sample: 1 1289

Included observations: 1289

Variable	Coefficient	Std. Error	t-Statistic	Prob.
C	0.905504	0.074175	12.20768	0.0000
FE	−0.249154	0.026625	−9.357891	0.0000
NW	−0.133535	0.037182	−3.591399	0.0003
UN	0.180204	0.036955	4.876316	0.0000
ED	0.099870	0.004812	20.75244	0.0000
EX	0.012760	0.001172	10.88907	0.0000

R-squared	0.345650	Mean dependent var		2.342416
Adjusted R-squared	0.343100	S.D. dependent var		0.586356
S.E. of regression	0.475237	Akaike info criterion		1.354639
Sum squared resid	289.7663	Schwarz criterion		1.378666
Log likelihood	−867.0651	Hannan-Quinn criter.		1.363658
F-statistic	135.5452	Durbin-Watson stat		1.942506
Prob(F-statistic)	0.000000			

a. How would you interpret these results?

b. Are the signs of the various coefficients in accord with prior expectations?

c. What is the semielasticity of the wage rate with respect to education? And with respect to years of experience?

d. Is it possible to compute the wage elasticity with respect to the dummy variables?[28]

e. Is the R^2 value given in the preceding table comparable with the R^2 value given in Table 4.1? Why or why not?

---

[28]On this, see Gujarati, D. (2015). *Econometrics by example* (2nd ed., pp. 60–61). London, England: Palgrave Macmillan.

Appendix 7A: Ridge Regression:
A Solution to Perfect Collinearity[29]

This topic is vast. We will present only the basic ideas underlying ridge regression. The important point is that instead of looking for a BLUE (best linear unbiased estimator) or BUE (best unbiased estimator), we should consider estimators that may be biased but have lower variances. In other words, there is a trade-off involved here. Therefore, we should consider the **minimum mean square error (MSE) estimator**, which strikes a balance between bias and minimum variance.

Let $\hat{\theta}$ be an estimator of the parameter θ. The MSE of $\hat{\theta}$ is defined as

$$\text{MSE}(\hat{\theta}) = E(\hat{\theta} - \theta)^2 \qquad (7A.1)$$

This is in contrast with the variance of $\hat{\theta}$, which is defined as

$$\text{var}(\hat{\theta}) = E[\hat{\theta} - E(\hat{\theta})]^2 \qquad (7A.2)$$

The difference between the two is that the variance of $\hat{\theta}$ is measured around its expected value, whereas MSE measures dispersion around the true value of the parameter.

The relationship between the two measures is as follows:

$$
\begin{aligned}
\text{MSE}(\hat{\theta}) &= E(\hat{\theta} - \theta)^2 \\
&= E[\hat{\theta} - E(\hat{\theta}) + E(\hat{\theta}) - \theta]^2 \\
&= E[\hat{\theta} - E(\hat{\theta})]^2 + E[E(\hat{\theta}) - \theta]^2 + 2E\{[\hat{\theta} - E(\hat{\theta})][E(\hat{\theta}) - \theta]\} \\
&= E[\hat{\theta} - E(\hat{\theta})]^2 + E[(\hat{\theta}) - \theta]^2 \\
&= E[\hat{\theta} - E(\hat{\theta})]^2 + [E(\hat{\theta}) - \theta]^2 \\
&= \text{var}(\hat{\theta}) + \text{bias}(\hat{\theta})^2 \qquad (7A.3)
\end{aligned}
$$

Note: The cross-product in Step (7A.2) is zero. Also note that θ and $E(\hat{\theta})$ are constants. As (7A.3) shows, MSE provides a trade-off between the variance and the bias of an estimator. Therefore, in some situations we may be able to accept a biased estimator if its variance is smaller than the variance of an unbiased estimator. This is the underlying principle of ridge regression, for ridge estimators try to balance the trade-off.

---

[29] The seminal article is by Hoerl, A., & Kennard, R. (1970). Ridge regression: Biased estimation for nonorthogonal problems. *Technometrics, 12*, 55–67.

Ridge regression estimators[30]

Recall that the usual OLS estimator b of B in the LRM

$$b = (X'X)^{-1} X'y \tag{7A.4}$$

$$\text{cov}(b) = \sigma^2 (X'X)^{-1} \tag{7A.5}$$

The diagonal elements of cov(b) give the variances of the elements of b. In cases of severe collinearity, these variances tend to be large, thus reducing the t values. Ridge regression is one method of "attenuating" these variances.

The simplest of the ridge estimators is called the **ordinary ridge regression (ORR)** and is defined as

$$
\begin{aligned}
b^R &= (X'X + kI)^{-1} X'y \\
&= (X'X + kI)^{-1} X'Xb
\end{aligned}
\tag{7A.6}
$$

The addition of kI to $X'X$ is to "regularize" the $X'X$ matrix so that it can be inverted even if it is near singular due to severe collinearity, where k is a positive number. The factor k is called the **shrinkage parameter**.

In (7A.6) b^R is a vector of ridge regression coefficients and b is the traditional OLS estimator.

Now

$$E(b^R) = (X'X + kI)^{-1} X'X E(b) \tag{7A.7}$$

which shows that b^R is a biased estimator, unless $k = 0$.

And

$$\text{cov}(b^R) = E(b^R - B)(b^R - B)' = \sigma^2 (X'X + kI)^{-1} X'X (X'X + kI)^{-1} \tag{7A.8}$$

It can be shown that the $[\text{cov}(b) - \text{cov}(b^R)]$ is PD (for $k > 0$), that is, the covariance matrix of the ridge estimator is smaller than the covariance matrix of the OLS estimator. This implies that

$$\text{var}(b_j^R) < \text{var}(b_j), \quad j = 1, 2, \ldots, k \tag{7A.9}$$

---

[30]The following discussion is based on Kmenta, J. (1986). *Elements of econometrics* (2nd ed., pp. 439–441). New York, NY: Macmillan. For a more technical discussion, see Fahrmeir, L., Kneib, T., Lang, S., & Marx, B. (2013). *Regression: Models, methods and applications* (pp. 203–208). Berlin, Germany: Springer-Verlag. This book also gives a concrete application. See also Ryan, T. P. (2009). *Modern regression methods* (2nd ed., chap. 12). New York, NY: Wiley.

The trade-off between bias and variance depends on the value of k—the larger the value of k, the larger the bias but the smaller the variance. Hoerl and Kennard (see Footnote 8) have shown that there exists a k such that

$$\text{tr MSE}(b^R) \leq \text{tr MSE}(b) \tag{7A.10}$$

A feature of ORR is that

$$b^{R\prime}b^R < b\prime b \tag{7A.11}$$

which shows that ORR pulls the OLS toward zero. That is why ORR belongs to a class of estimators called *shrinkage estimators*.

Until recently, ridge estimators were rarely used because k is rarely known and because the traditional distribution theory is not applicable to ridge regression. However, ridge regression has renewed interest in applications with a large number of regressors but relatively small sample size.[31]

Several software packages have modules to estimate ridge regressions.

Appendix 7B: Specification Errors

Underfitting a model (omitting relevant variables)

Consider the standard LRM:

$$y = XB + u \tag{7B.1}$$

where y is $n \times 1$, X is $n \times k$, and u is $n \times 1$

$$y = X_1 B_1 + X_2 B_2 + u \tag{7B.2}$$

where X_1 is $n \times k_1$, X_2 is $n \times k_2$, and $(k_1 + k_2) = k$.

Assume that we erroneously exclude X_2 from (7B.1) and estimate the model

$$y = X_1 B_1^* + v \tag{7B.3}$$

Using the standard OLS formula, we estimate (7B.3) and obtain

$$b_1^* = \left(X_1'X_1 \right)^{-1} X_1'y \tag{7B.4}$$

---

[31]See Hastie, T. J., Tibshirain, R. J., & Friedman, J. (2009). *The elements of statistical learning*. Berlin, Germany: Springer.

Substituting for y from (7B.2), we obtain

$$b_1^* = \left(X_1'X_1\right)^{-1} X_1'[X_1B_1 + X_2B_2 + u]$$
$$= \left(X_1'X_1\right)^{-1} X_1'X_1B_1 + (X_1'X_1)^{-1} X_1'X_2B_2 + \left(X_1'X_1\right)^{-1} X_1'u \qquad (7B.5)$$

Now,

$$E(b_1^*) = B_1 + (X_1'X_1)^{-1} X_1'X_2B_2 \qquad (7B.6)$$
$$= B_1 + AB_2$$

since the X matrices are nonstochastic and $E(u)=0$.
The matrix $A = (X_1'X_1)^{-1} X_1'X_2$ is called the **alias matrix**.

$$\text{cov}\left(b_1^*\right) = \text{cov}[(X_1'X_1)^{-1} X_1'y]$$
$$= (X_1'X_1)^{-1} X_1'(\sigma^2 I)X_1(X_1'X_1)^{-1}$$
$$= \sigma^2 (X_1'X_1)^{-1} \qquad (7B.7)$$

As (7B.6) shows, b_1 is biased and the bias depends on the values of the regressors in both the matrices. If you look closely at the alias matrix A, it is simply the regression of the regressors included in the X_2 matrix on the regressors in the X_1 the matrix. And unless $X_1'X_2 = 0$, that is, the columns in X_1 are orthogonal to the columns in X_2, or $B_2 = 0$, b_1 will remain biased. Not only that, it is also inconsistent.

It can be shown that[32]

$$\text{cov}(b_1) - \text{cov}\left(b_1^*\right) = \sigma^2 AB^{-1}A' \qquad (7B.8)$$

which is a PD matrix, where $A = (X_1'X_1)^{-1} X_1'X_2$ and $B = X_2'X_2 - X_2'X_1A$.

As a result, $\text{var}(b_j) > \text{var}(b_j^*)$, which means that underfitting a model reduces the variance of bs. In addition, it can be shown that underfitting a model also biases the error variance of the error term u. As we know, if there is no specification error, the estimate of the error variance is given by

$$S^2 = e'e / (n-k) \qquad (7B.9)$$

---

[32]For proofs, see Rencher, A. C. (2000). *Linear models in statistics* (pp. 155–157). New York, NY: Wiley Inter-Science.

but in case of underfitting, the error variance is given by

$$S_1^2 = \frac{e^{*'}e^*}{n - k_1} \tag{7B.10}$$

where e^* is the residual from the underfitted model and $E(S_1^2)$ is not equal to the true σ^2, whereas $E(S^2) = \sigma^2$.[33]

Overfitting a model (adding superfluous variables)

We now consider the case where the true model is

$$y = X_1 B_1 + v \tag{7B.11}$$

but we estimate

$$y = X_1 B_1 + X_2 B_2 + u = XB + v \tag{7B.12}$$

even though $B_2 = 0$.
Therefore, we obtain

$$b = (X'X)^{-1} X'y \tag{7B.13}$$

Now,

$$\begin{aligned} E(b) &= (X'X)^{-1} X'(X_1 B_1 + v) \\ &= (X'X)^{-1} X'X_1 B_1 \end{aligned} \tag{7B.14}$$

We can express the matrix X_1 as follows:

$$X_1 = [X_1 X_2] \begin{bmatrix} I_{k_1} \\ 0 \end{bmatrix} = X \begin{bmatrix} I_{k_1} \\ 0 \end{bmatrix}$$

where I_{k_1} is an identity matrix of order k_1 and 0 is a $(k_1 \times k_2)$ matrix of zeros.

---

[33]See Rencher, A. C. (2000). *Linear models in statistics* (p. 157). New York, NY: Wiley Inter-Science.

As a result,

$$E(b) = (X'X)^{-1} X'X \begin{bmatrix} I_{k_1} \\ 0 \end{bmatrix} B_1 = \begin{bmatrix} B_1 \\ 0 \end{bmatrix} \tag{7B.15}$$

This shows that B_1 is unbiased even though redundant variables are included in the model.

It can be shown that S^2, the estimator of the error variance, σ^2, is still unbiased and so we can continue with statistical inference in the usual manner. In sum, in this case, the OLS estimators are BLUE. But it can be shown that the inclusion of redundant regressors lowers the precision of the estimators relative to when they are not included in the model. This is because additional degrees of freedom are consumed in the estimation of redundant parameters in B_2.

APPENDIX A: BASICS OF MATRIX ALGEBRA

In what follows, we will denote matrices by bold capital letters and vectors by bold lowercase letters.

A.1 Definitions

A.1.1 Matrix

A matrix is a rectangular array of real numbers arranged in rows and columns. More formally, a matrix of **order**, or **dimension**, m by n (written as $m \times n$) is a set of $m \times n$ real numbers arranged in m rows and n columns, as the matrix A below.

$$A_{m \times n} = [a_{ij}] = \begin{pmatrix} a_{11} & \cdots & a_{1n} \\ \vdots & \ddots & \vdots \\ a_{m1} & \cdots & a_{mn} \end{pmatrix} \tag{A.1}$$

where a_{ij} is the element appearing in the ith row and jth column and where $[a_{ij}]$ is a shorthand expression for the matrix A whose typical element is a_{ij}. The subscript on the matrix A indicates the dimension of the matrix, that is, the number of rows and columns, m and n, respectively, in this case. Of course, in practice m and n will have numerical values. Thus, a matrix $B_{2 \times 3}$ means this matrix has 2 rows and 3 columns.

A.1.2 Scalar

A scalar is single real number. Alternatively, a scalar is a 1×1 matrix.

A.1.3 Column Vector

A matrix consisting of m rows and one column ($m \times 1$) is called a **column vector**. Thus,

$$x_{3 \times 1} = \begin{bmatrix} 3 \\ 4 \\ 7 \end{bmatrix} \tag{A.2}$$

is an example of a 3×1 column vector.

A.1.4 Row Vector

A matrix consisting of only 1 row and n columns is called a **row vector**. Thus,

$$x_{1\times4} = \begin{bmatrix} 1 & 3 & 8 & -4 \end{bmatrix} \tag{A.3}$$

is an example of a 1×4 row vector.

A.1.5 Transposition of a Matrix

The transpose of an $m \times n$ matrix A, denoted by A' (read as A prime or A transpose) is an $n \times m$ matrix obtained by interchanging the rows and columns of A, that is, the ith row of A becomes the ith column of A'. Thus,

$$A_{4\times2} = \begin{bmatrix} 5 & 9 \\ 3 & 11 \\ 8 & 2 \\ -7 & 6 \end{bmatrix} \quad A'_{2\times4} = \begin{bmatrix} 5 & 3 & 8 & -7 \\ 9 & 11 & 2 & 6 \end{bmatrix} \tag{A.4}$$

As you can see, the transpose of a matrix is obtained by rewriting its columns as rows. Note that some textbooks use A^{T} to denote the transpose of A.

A.1.6 Transpose of Vectors

The transpose of a row vector is a column vector, and vice versa. Usually, a vector with a prime sign (') or a transpose sign (T) is a row vector and a vector without such a sign is a column vector.

A.1.7 Submatrix

Given an $m \times n$ matrix A, if we delete all but r rows and s columns, the resulting $r \times s$ matrix is called a submatrix of A. Thus, if

$$A_{3\times3} = \begin{bmatrix} 5 & 6 & 7 \\ 10 & 2 & 3 \\ 4 & 8 & 9 \end{bmatrix}$$

and if we delete the third row and the third column of this matrix, we obtain

$$B_{2\times2} = \begin{bmatrix} 5 & 6 \\ 10 & 2 \end{bmatrix} \tag{A.5}$$

which is a submatrix of A of order 2×2.

A.2 Types of Matrices

A.2.1 Square Matrix

A matrix that has the same number of rows as columns is called a **square matrix**. The A and B matrices above are examples of square matrices.

A.2.2 Diagonal Matrix

A square matrix with at least one nonzero element on the main diagonal (running from the upper-left corner to the lower right-hand corner) and zero elsewhere is called a **diagonal matrix**. Another way of expressing this is that a square matrix with off-diagonal elements all zero is called a diagonal matrix. An example is

$$A_{3\times3} = \begin{bmatrix} -4 & 0 & 0 \\ 0 & 7 & 0 \\ 0 & 0 & 9 \end{bmatrix} \tag{A.6}$$

A.2.3 Identity or Unit Matrix

A diagonal matrix whose diagonal elements are all 1 is called an **identity**, or **unit**, **matrix** and is denoted by I. The following is an example of an identity matrix.

$$I_{3\times3} = \begin{bmatrix} 1 & 0 & 0 \\ 0 & 1 & 0 \\ 0 & 0 & 1 \end{bmatrix} \tag{A.7}$$

A.2.4 Scalar Matrix

A diagonal matrix whose diagonal elements are equal is called a **scalar matrix**. An example is the variance–covariance matrix of the error term u in the classical linear regression model, namely,

$$\text{var-cov}(u) = \begin{bmatrix} \sigma^2 & 0 & 0 & 0 & 0 \\ 0 & \sigma^2 & 0 & 0 & 0 \\ 0 & 0 & \sigma^2 & 0 & 0 \\ \vdots & & & \ddots & \\ 0 & 0 & 0 & 0 & \sigma^2 \end{bmatrix} = \sigma^2 \begin{bmatrix} 1 & 0 & 0 & 0 & 0 \\ 0 & 1 & 0 & 0 & 0 \\ 0 & 0 & 1 & 0 & 0 \\ \vdots & & & \ddots & \\ 0 & 0 & 0 & 0 & 1 \end{bmatrix} \tag{A.8}$$

As you can see, an identity matrix is a special kind of a scalar matrix.

A.2.5 Symmetric Matrix

If a square matrix A is the same as its transpose, it is called a symmetric matrix, that is, if $A'=A$. In this case, the element a_{ij} of matrix A is equal to the element a_{ji} of A'. An example is the variance–covariance matrix of the error term u in the classical linear regression model. See, for example, Equation (A.1).

A.2.6 Null Matrix

A matrix whose elements are all zero is called a null matrix and is denoted by **0**.

A.2.7 Null Vector

A row or column vector whose elements are all zero is called a **null vector** and is also denoted by **0**.

A.2.8 Equal Matrices

Two matrices A and B are said to be equal if they are of the same order and their corresponding elements are equal, that is, $a_{ij}=b_{ij}$ for all i and j. As an example,

$$A_{3\times3} = \begin{bmatrix} 5 & 7 & 0 \\ 0 & -2 & 3 \\ 4 & 1 & 8 \end{bmatrix} \text{ and } B_{3\times3} = \begin{bmatrix} 5 & 7 & 0 \\ 0 & -2 & 3 \\ 4 & 1 & 8 \end{bmatrix} \tag{A.9}$$

are equal, because $A=B$.

A.2.9 Upper Triangular Matrix

A square matrix with zeros below the main diagonal.

A.2.10 Lower Triangular Matrix

A square matrix with zeros above the main diagonal.

A.2.11 Idempotent Matrix

A square matrix A such that $A^n=A$ is called an **idempotent (of the same power) matrix**. That is, no matter how many times you multiply the matrix by itself, the resulting product matrix is the same as the original matrix A. Thus, $A^2=A$, $A^3=A$, and so on. This is because all the eigenvalues of such a matrix are 1 or 0. Note that $A^0 = I$. On eigenvalues, see Section A.11.

A.3 Matrix Operations

A.3.1 Matrix Addition

Let $A = [a_{ij}]$ and $B = [b_{ij}]$. If both of these matrices are of the same order, we define matrix addition as

$$A + B = C \tag{A.10}$$

where C is of the same order as A and B and is obtained by taking the sums $c_{ij} = a_{ij} + b_{ij}$ for all i and j, that is, C is obtained by adding the corresponding elements of A and B. If this can be done, A and B are said to be **conformable** for addition. For instance,

$$A = \begin{bmatrix} 3 & -2 & 4 & 6 \\ 8 & 7 & 4 & 11 \end{bmatrix} \text{ and } B = \begin{bmatrix} 5 & 8 & 4 & -3 \\ 9 & 1 & 7 & 6 \end{bmatrix}$$

$$C = A + B = \begin{bmatrix} 8 & 6 & 8 & 3 \\ 17 & 8 & 11 & 17 \end{bmatrix}$$

A.3.2 Matrix Subtraction

Matrix subtraction follows the same principle as matrix addition except that $C = A - B$. For example, using the matrices A and B given above, we find that

$$C = A - B = \begin{bmatrix} -2 & -10 & 0 & 9 \\ -1 & 6 & -3 & 5 \end{bmatrix}$$

A.3.3 Scalar Multiplication

To multiply a matrix A by a scalar c (a real number), we multiply each element of the matrix by c. Therefore, $cA = Ac = [ca_{ij}]$. Thus, if $c = 2$ and $A = \begin{bmatrix} -8 & 7 \\ 6 & 5 \end{bmatrix}$, then

$$2A = \begin{bmatrix} -16 & 14 \\ 12 & 10 \end{bmatrix}$$

A.3.4 Matrix Multiplication

Let A be $m \times n$ and B be $n \times p$. Then the product AB, in that order, is defined to be a new matrix C of order $m \times p$, such that

$$c_{ij} = \sum_{k=1}^{n} a_{ik} b_{kj}, \quad i=1,2,\ldots,m \text{ and } j=1,2,\ldots,p \qquad (A.11)$$

That is, the element in the ith row and the jth column of C is obtained by multiplying the elements of the ith row of A by the corresponding elements of the jth column of B and summing over terms. This is known as the **row by column rule of multiplication**. Thus, to obtain c_{11}, the element in the first row and first column of C, we multiply the elements in the first row of A by the corresponding elements in the first column of B and sum over all terms, and so on.

Note that for multiplication to exist, matrices A and B *must be conformable with respect to multiplication*, that is, the number of columns in A must be equal to the rows in B. For example,

$$A_{2\times3} = \begin{bmatrix} 3 & 4 & 7 \\ 5 & 6 & 1 \end{bmatrix} \text{ and } B_{3\times2} = \begin{bmatrix} 2 & 1 \\ 3 & 5 \\ 6 & 2 \end{bmatrix}$$

$$AB = C_{2\times2} = \begin{bmatrix} (3\times2)+(4\times3)+(7\times6) & (3\times1)+(4\times5)+(7\times2) \\ (5\times2)+(6\times3)+(1\times6) & (5\times1)+(6\times5)+(1\times2) \end{bmatrix}$$

$$= \begin{bmatrix} 60 & 37 \\ 34 & 37 \end{bmatrix}$$

Note that in this example the resulting C matrix is of order 2×2.

In this example, is the product BA defined? If so, what is the dimension of the resulting product matrix?

A.3.4.1 Properties of Matrix Multiplication

1. Matrix multiplication is not necessarily **commutative**, that is, in general, $AB \neq BA$. The order in which matrices are multiplied is very important. AB means A is **postmultiplied** by B or B is **premultiplied** by A.

2. Even if AB and BA exist, the resulting matrices may not be of the same order. For example, if A is $m \times n$ and B is $n \times m$, AB is $m \times m$, but BA is $n \times n$, and hence of a different order.

3. Even if A and B are both square matrices, so that AB and BA are both defined, the resulting matrices will not necessarily be equal. For example, consider

$$A = \begin{bmatrix} 4 & 7 \\ 3 & 2 \end{bmatrix} \text{ and } B = \begin{bmatrix} 1 & 5 \\ 6 & 8 \end{bmatrix}; \text{ then } AB = \begin{bmatrix} 46 & 76 \\ 15 & 31 \end{bmatrix} \text{ and } BA = \begin{bmatrix} 19 & 17 \\ 48 & 58 \end{bmatrix}$$

Thus, $AB \neq BA$. But if both A and B are **identity matrices**, then $AB = BA$.

4. A row vector multiplied by a column vector is a scalar. As an example, consider the OLS residuals e_1, e_2, ..., e_n. Letting e be a column vector of the residuals and e' the row vector of the residuals, we can see that

$$e'e = \begin{bmatrix} e_1 & e_2 & e_3 & \cdots & e_n \end{bmatrix} \begin{bmatrix} e_1 \\ e_2 \\ e_3 \\ \vdots \\ e_n \end{bmatrix} = e_1^2 + e_2^2 + e_3^2 + \cdots + e_n^2 = \sum_1^n e_i^2 \quad \text{(A.12)}$$

which is a scalar, or 1×1 matrix. Note that e' is $1 \times n$, but e is $n \times 1$.

5. But see what happens if we multiply an $n \times 1$ column vector by a $1 \times n$ row vector. We get an $n \times n$ matrix:

$$ee' = \begin{bmatrix} e_1 \\ e_2 \\ e_3 \\ \vdots \\ e_n \end{bmatrix} \begin{bmatrix} e_1 & e_2 & e_3 & \cdots & e_n \end{bmatrix} = \begin{pmatrix} e_1^2 & \cdots & e_1 e_n \\ \vdots & \ddots & \vdots \\ e_n e_1 & \cdots & e_n^2 \end{pmatrix} \quad \text{(A.13)}$$

6. A matrix multiplied by a column vector is a column vector. If $A_{m \times n}$ and $B_{n \times 1}$, then AB is $(m \times 1)$, a column vector.

7. A row vector postmultiplied by a matrix is a row vector.

8. Matrix multiplication is **associative**, that is, $(ABC) = A(BC)$, where A is $m \times n$, B is $n \times p$, and C is $p \times k$.

9. Matrix multiplication is **distributive** with respect to addition, thus

$$A(B+C)=AB+AC \text{ and } (B+C)A=BA+CA$$

A.4 Matrix Transposition

1. The transpose of a transposed matrix is the original matrix: $(A')'=A$.

2. The transpose of the sum of two (conformable) matrices is the sum of their respective transposes. Thus, if A and B are conformable for addition, then

$$C=A+B \text{ and } C' =(A+B)' = A' + B'$$

3. If the product AB is defined, then $(AB)'=B'A'$. This can be generalized. If the product $(ABCD)$ is defined, then $(ABCD)'=D'C'B'A'$.

4. The transpose of an identity matrix I is the identity matrix itself, that is, $I'=I$.

5. The transpose of a scalar is the scalar itself. Thus, if γ is a scalar, $\gamma' = \gamma$.

6. The transpose of $(\gamma'A)'$ is $\gamma A'$. This is because $(\gamma'A)'=A'\gamma''=A'\gamma=\gamma A'$.

7. If A is a square matrix such that $A=A'$, then A is a symmetric matrix.

A.5 Matrix Inversion

An inverse of a square matrix A, denoted by A^{-1} (read as A inverse), if it exists, is a unique square matrix such that

$$AA^{-1} = A^{-1}A = I$$

where I is an identity matrix whose order is the same as that of A. For example,

$$A=\begin{bmatrix} 2 & 4 \\ 6 & 8 \end{bmatrix} A^{-1} = \begin{bmatrix} -1 & \dfrac{1}{2} \\ \dfrac{6}{8} & -\dfrac{1}{4} \end{bmatrix} AA^{-1} = \begin{bmatrix} 1 & 0 \\ 0 & 1 \end{bmatrix} = I$$

Properties of matrix inversion are given in Section A.8.

A.6 Determinants

To every square matrix A, there corresponds a number (scalar) known as the determinant of the matrix, which is denoted by det A or by the symbol $|A|$, the two parallel lines that enclose A meaning "the determinant of" and not the usual symbol for the absolute value. Note that a matrix per se has no numerical value, but the determinant of a matrix is a number.

A.6.1 Evaluation of a Determinant

The process of finding the value of a determinant is known as **evaluation**, **expansion**, or **reduction** of the determinant. This is accomplished by manipulating the entries of the matrix in a well-defined manner.

Evaluation of a 2×2 determinant:

If

$$A = \begin{bmatrix} a_{11} & a_{12} \\ a_{21} & a_{22} \end{bmatrix},$$

its determinant is evaluated as

$$|A| = \begin{vmatrix} a_{11} & a_{12} \\ a_{21} & a_{22} \end{vmatrix} = a_{11}a_{22} - a_{12}a_{21}$$

which is obtained by cross-multiplying the elements on the main diagonal and subtracting from this the cross-multiplication of the elements on the other diagonal of matrix A, as shown by the arrows.

Evaluation of a 3×3 determinant:

If

$$A = \begin{bmatrix} a_{11} & a_{12} & a_{13} \\ a_{21} & a_{22} & a_{23} \\ a_{31} & a_{32} & a_{33} \end{bmatrix}$$

$$|A| = a_{11}a_{22}a_{33} - a_{11}a_{23}a_{32} + a_{12}a_{23}a_{31} - a_{12}a_{21}a_{33} + a_{13}a_{21}a_{32} - a_{13}a_{22}a_{31}$$

An examination of the preceding pattern is not arbitrary. Note the following:

1. Each term in the expansion of the determinant contains one and only one element from each row and each column.

2. The number of elements in each term is the same as the number of rows or columns in the matrix: A 2×2 determinant has two elements in each term of expansion, a 3×3 determinant has three elements in each term of expansion, and so on.

3. The general rule is that the determinant of order $n \times n$ has $n! = n \times (n-1) \times \cdots \times 3 \times 2 \times 1$ terms in the expansion where $n!$ is called "n factorial." Thus, a 6×6 determinant will have $6 \times 5 \times 4 \times 3 \times 2 \times 1 = 720$ terms in the expansion.

4. The terms in the expansion alternate in sign from $+$ to $-$.

A.6.2 Properties of Determinants

1. A matrix whose determinant is zero is called a **singular matrix**, whereas a matrix with a nonzero determinant is called a **nonsingular matrix**. The inverse of a singular matrix does not exist.

2. If all the elements of any row of A are zero, its determinant is zero.

3. $|A'| = |A|$, that is, the determinants of A and A transpose are the same.

4. Interchanging any two rows or any two columns of A changes the sign of $|A|$.

5. If every element of a row or a column of A is multiplied by a scalar λ, then $|A|$ is multiplied by λ.

6. If two rows or columns of a matrix are identical, its determinant is zero.

7. If one row or a column is a multiple of another row or column of that matrix, its determinant is zero.

8. $|AB| = |A| \, |B|$, that is, the determinant of the product of two matrices is the product of their individual determinants.

9. If $A = \mathrm{diag}(a_1, a_2, \ldots, a_n)$, and all off-diagonal elements are zero, $|A| = a_1 \cdot a_2 \cdot a_3 \cdots a_n$, that is, the product of all the diagonal elements.

A.7 Rank of a Matrix

The rank of a matrix is the maximum number of linearly independent rows, which is the same as the maximum number of linearly independent columns. Thus, the rank of a matrix is equal to that of its transpose. If a matrix has m rows and n columns, with $m \leq n$, then the rank is $\leq m$; if $m > n$, then the rank is $\leq n$. If the rank is equal to the smaller of m and n, then the matrix is of **full rank**. The rank of a matrix can also be defined as the order of the largest square submatrix whose determinant is not zero.

For example, for the matrix

$$A = \begin{bmatrix} 3 & 6 & 6 \\ 0 & 4 & 5 \\ 3 & 2 & 1 \end{bmatrix}$$

its determinant is zero because Row 2+Row 3=Row 1, hence it is a singular matrix. Hence, although its order is 3×3, its rank is less than 3. Actually it is 2, because we can find a 2×2 submatrix whose determinant is not zero. For instance, if we delete the first row and the first column of A, we obtain

$$B = \begin{bmatrix} 4 & 5 \\ 2 & 1 \end{bmatrix}$$

The determinant of this matrix is −6, which is nonzero, thus establishing that the rank of the matrix A is in fact 2.

A.7.1 Minor

If the ith row and the jth column of an $n \times n$ matrix A are deleted, the determinant of the resulting submatrix is called the **minor** of the element a_{ij}, the element at the intersection of the ith row and jth column, and is denoted by $|M_{ij}|$.

As an example, consider the following matrix:

$$A = \begin{pmatrix} a_{11} & a_{12} & a_{13} \\ a_{21} & a_{22} & a_{23} \\ a_{31} & a_{32} & a_{33} \end{pmatrix}$$

The minor of a_{11} is $|M_{11}| = \begin{bmatrix} a_{22} & a_{23} \\ a_{32} & a_{33} \end{bmatrix} = a_{22}a_{33} - a_{23}a_{32}$.

The minors of other elements of A can be found similarly.

A.7.2 Cofactor

The cofactor of the element a_{ij} of an $n \times n$ matrix A, denoted by c_{ij}, is defined as $c_{ij} = (-1)^{i+j} |M_{ij}|$. In words, a cofactor is a **signed minor**, the sign being positive if $i+j$ is even and being negative if $i+j$ is odd. For instance, the cofactor of the element a_{11} of a 3×3 matrix A given previously is $a_{22}a_{33} - a_{23}a_{32}$, whereas the cofactor of the element a_{21} is $-(a_{12}a_{33} - a_{13}a_{32})$, since the sum of the subscripts 2 and 1 is 3, which is an odd number.

A.7.3 Cofactor Matrix

Replacing the elements of a_{ij} of a matrix A by their cofactors, we obtain a matrix known as the **cofactor matrix** of A, denoted by (cof A).

A.7.4 Adjoint Matrix

The adjoint matrix, written as (adj A), is the transpose of the cofactor matrix, that is, (adj A) = (cof A)'.

A.8 Finding the Inverse of a Square Matrix

If A is square and nonsingular, that is, $|A| \neq 0$, its inverse can be found as follows:

$$A^{-1} = \frac{1}{|A|}(\text{adj } A)$$

The steps involved in the computation of the inverse are as follows:

1. Find the determinant of A; if it is nonzero, proceed to Step 2.

2. Replace element a_{ij} of A by its cofactors to obtain the cofactor matrix.

3. Transpose the cofactor matrix to obtain the adjoint matrix.

4. Divide each element of the adjoint matrix by $|A|$.

5. If you multiply A by A^{-1}, you should get an identity matrix.

Example: Consider the following matrix:

$$A = \begin{bmatrix} 1 & 2 & 3 \\ 5 & 7 & 4 \\ 2 & 1 & 3 \end{bmatrix}$$

It is left for the reader to verify that the inverse of this matrix is as follows:

$$A^{-1} = -\frac{1}{24}\begin{bmatrix} 17 & -3 & -13 \\ -7 & -3 & 11 \\ -9 & 3 & -3 \end{bmatrix}$$

A.8.1 Properties of Inverse Matrices

1. If A and B are square matrices of the same size such that both are nonsingular, then $(AB)^{-1} = B^{-1}A^{-1}$.

2. If A is nonsingular, then $(A')^{-1} = (A^{-1})'$, that is, the inverse of the transposed matrix is the transpose of its inverse. That is, the inverse symbol (-1) and the transpose symbol (') are interchangeable.

3. If A is nonsingular, $|A^{-1}| = \dfrac{1}{|A|}$.

4. $(A^{-1})^{-1} = A$. That is, the inverse of a nonsingular matrix is the original matrix.

5. If A is nonsingular, the system of equation $Ax = c$ has the unique solution $x = A^{-1}c$, where x and c are appropriately defined vectors.

A.9 Trace of a Square Matrix

The trace of a square matrix A, denoted by tr(A), is defined as the sum of its diagonal elements, that is,

$$\text{tr}(A) = a_{11} + a_{22} + \cdots + a_{nn}$$

Some of the properties of the trace are as follows:

1. tr(k) = k, where k is a constant.

2. tr(A') = tr(A), that is, the trace of a transposed matrix is equal to the trace of the original matrix.

A.10 Quadratic Forms and Definite Matrices

If A is a symmetric matrix and x is a vector, the product

$$x'Ax = \sum_{i}^{n}\sum_{j}^{n} a_{ij}x_i x_j$$

is called a **quadratic form**.

Note that the symmetry of A means $a_{ij}x_ix_j = a_{ji}x_ix_j$.

A quadratic form is called **positive definite (PD)** if $x'Ax > 0$ for all non-null vectors. It is called **positive semidefinite (PSD)** if $x'Ax \geq 0$. It is called **negative definite (ND)** if $x'Ax < 0$ and **negative semidefinite (NSD)** if $x'Ax \leq 0$. These matrices play an important role in statistics. Note that *a quadratic form is a scalar*, hence it equals its trace, that is, $x'Ax = \text{tr}(x'Ax) = \text{tr}(Axx')$.

A.10.1 Some Properties of Quadratic Forms

1. If A is an $n \times n$ PD matrix, then $|A| > 0$, rank of $A = n$ and A is nonsingular.

2. If A is PD, then A^{-1} is also PD.

3. If A is PD, there exists a nonsingular matrix P such that $PAP' = I$ and $P'P = A^{-1}$.

4. If A and B are symmetric and $(A - B)$ is PD, then $|A| \geq |B|$ and $x'Ax \geq x'Bx$ for all x.

A.10.2 Mean and Variance of Quadratic Forms[1]

If x is a random vector with mean μ and covariance Σ and if A is a symmetric matrix of constants, then

$$E(x'Ax) = \text{tr}(A\Sigma) + \mu'A\mu$$

and its variance is

$$\text{var}(x'Ax) = 2\text{tr}\left[(A\Sigma)^2\right] + 4\mu'A\Sigma A\mu$$

If $X \sim N(0, \Sigma)$, then $\text{var}(x'Ax) = 2\text{tr}(A\Sigma)^2$.

A.11 Eigenvalues and Eigenvectors

For every square matrix A, we can find a scalar λ and a nonzero vector x such that

$$Ax = \lambda x \tag{A.14}$$

In this equation, λ is called an **eigenvalue**, also known as a **characteristic root**, and x is known as an **eigenvector**, also known as a **characteristic vector**.

---

[1]For proofs, see Searle, S. R. (1971). *Linear models* (pp. 55–57). New York, NY: Wiley.

To find λ and x of A, we can write (A.14) as

$$(A - \lambda I)x = 0 \tag{A.15}$$

If the matrix $(A - \lambda I)$ is nonsingular, the only solution to (A.15) is $x = 0$. However, a nonzero solution is possible if $(A - \lambda I)$ is singular, which means the determinant $|(A - \lambda I)| = 0$. In this situation, we obtain a polynomial of order n in λ, which is known as the **characteristic polynomial** in A and the resulting equation is called the **characteristic equation**. The roots of the characteristic equation are called characteristic roots or eigenvalues. Corresponding to each λ_i we will have a characteristic vector x_i. In all, there will be n characteristic roots and characteristic vectors. It may be noted that not all characteristic roots will be distinct.

Example: Consider the following matrix

$$A = \begin{pmatrix} 2 & 2 \\ 2 & -1 \end{pmatrix} \tag{A.16}$$

Using this matrix and (A.15), we obtain the following determinant:

$$|A - \lambda I| = \begin{vmatrix} 2 - \lambda & 2 \\ 2 & -1 - \lambda \end{vmatrix} = \lambda^2 - \lambda - 6 = 0 \tag{A.17}$$

Since this is a polynomial of second degree, we will obtain two characteristic roots, namely, $\lambda_1 = 3$ and $\lambda_2 = -2$.

Using the first root and (A.14), we obtain

$$\begin{bmatrix} 2 & -3 & 2 \\ 2 & -1 & -3 \end{bmatrix} \begin{bmatrix} x_1 \\ x_2 \end{bmatrix} = \begin{bmatrix} -1 & 2 \\ 2 & -4 \end{bmatrix} \begin{bmatrix} x_1 \\ x_2 \end{bmatrix} = \begin{bmatrix} 0 \\ 0 \end{bmatrix} \tag{A.18}$$

Since the two rows of this matrix are linearly dependent, there is an infinite number of solutions, which can be expressed by the equation $x_1 = 2x_2$. To get a unique solution, we **normalize** the solution by imposing the restriction that $x_1^2 + x_2^2 = 1$, which yields $x_1^2 + x_2^2 = (2x_2)^2 + x_2^2 = 5x_2^2 = 1$. Taking the positive square root of this, we obtain $x_2 = \dfrac{1}{\sqrt{5}}$. Since $x_1 = 2x_2 = \dfrac{2}{\sqrt{5}}$, the first characteristic vector is $v_1 = \begin{bmatrix} 2/\sqrt{5} \\ 1/\sqrt{5} \end{bmatrix}$. The reader can verify that corresponding to the eigenvalue of -2, we obtain the second characteristic vector as $v_2 = \begin{bmatrix} -1/\sqrt{5} \\ 2/\sqrt{5} \end{bmatrix}$.

Note that $v_1'v_1 = 1$ and $v_1'v_2 = 0$. These two important properties can be generalized to as many as n characteristic vectors. When the scalar product of two vectors is zero, we call them **orthogonal vectors**, that is, they are perpendicular to each other. The other property of the scalar product of a characteristic root with itself is unity is because of the normalization rule. As a result of these two properties, the characteristic vectors of a matrix form what is called a set of **orthonormal vectors**.

The knowledge of characteristic roots and characteristic vectors is quite useful in determining the properties of quadratic forms, more specifically the following.

1. The quadratic form $x'Ax$ is positive (negative) definite, if and only if *every* characteristic root of A is positive (negative).

2. $x'Ax$ is positive (negative) semidefinite, if and only if all *characteristic roots* of A are nonnegative (nonpositive) and *at least* one root is zero.

3. $x'Ax$ is indefinite, if and only if some of the characteristic roots of A are positive and some are negative.

The knowledge of characteristic roots and vectors is also useful in the study of symmetric matrices. Some examples are as follows:

1. For any symmetric matrix A, there exists an *orthogonal* matrix H such that $H'AH=Z$, where Z is a diagonal matrix whose diagonal elements are the characteristic roots of A. Note that $H'H=I$.

2. The rank of a square matrix is equal to the number of its nonzero characteristic roots.

3. For any matrices X and Y, not necessarily square, the nonzero characteristic roots of XY and YX are the same, whenever both these product matrices are defined.

4. If A and B are symmetric matrices of the same size, then they both can be diagonalized by the same orthogonal matrix if and only if $AB=BA$.

5. The determinant of a square matrix is the product of its characteristic roots.

6. The trace of a square matrix is the sum of its characteristic roots.

A.12 Vector and Matrix Differentiation

If $y = f(x)$, a function of the variables $x_1, x_2, \ldots, x_k$, in $x = (x_1, x_2, \ldots, x_k)'$, and let $\partial y / \partial x_1, \partial y / \partial x_2, \ldots, \partial y / \partial x_k$ be the partial derivatives, then we define $\partial y / \partial x$ as

$$\frac{\partial y}{\partial x} = \begin{bmatrix} \partial y / \partial x_1 \\ \partial y / \partial x_2 \\ \vdots \\ \partial y / \partial x_k \end{bmatrix}$$

which we can also write as

$$\frac{\partial y}{\partial x'} = \left(\frac{\partial y}{\partial x_1}, \frac{\partial y}{\partial x_2}, \cdots, \frac{\partial y}{\partial x_n} \right)$$

which is the transpose of $\partial y / \partial x$.

If $a' = \begin{bmatrix} a_1 & a_2 & \cdots & a_k \end{bmatrix}$ is a row vector of numbers and $x = \begin{bmatrix} x_1 \\ x_2 \\ \vdots \\ x_k \end{bmatrix}$ is a col-

umn vector of the variables $x_1, x_2, \ldots, x_k$, and if $y = a'x = x'a$, then

$$\frac{\partial a'x}{\partial x} = \frac{\partial x'a}{\partial x} = a = \begin{bmatrix} a_1 \\ a_2 \\ \vdots \\ a_k \end{bmatrix}$$

A.12.1 Some Differentiation Rules

Let A be a matrix and a, x, y be vectors. Assuming the usual matrix rules about multiplication, division (inverse), and the like, the following rules exist:

1. $\dfrac{\partial y'x}{\partial x} = y$

2. $\dfrac{\partial x'Ax}{\partial x} = (A + A')x$

$\qquad = 2Ax = 2A'x \quad$ if A is symmetric

3. $\dfrac{\partial Ax}{\partial x} = A'$

4. $\dfrac{\partial Ax}{\partial x'} = A$

A.12.2 The Hessian Matrix

Let $x = (x_1, x_2, \ldots, x_n)'$ be an $n \times 1$ vector and $f(x)$ a real function differentiable with respect to x_i, an element of the x vector. Let $m(x) = \dfrac{\partial f(x)}{\partial x}$ represent first derivatives. Let $H(x)$ represent the first derivative of $m(x)$, that is, the second derivative of the original function $f(x)$ as follows:

$$H(x) = \frac{\partial m(x)}{\partial x'} = \begin{bmatrix} \dfrac{\partial m_1(x)}{\partial x_1} & \cdots & \dfrac{\partial m_1(x)}{\partial x_n} \\ \vdots & \ddots & \vdots \\ \dfrac{\partial m_n(x)}{\partial x_1} & \cdots & \dfrac{\partial m_n(x)}{\partial x_n} \end{bmatrix}$$

$$= \begin{bmatrix} \dfrac{\partial^2 f(x)}{\partial x_1 \partial x_1} & \cdots & \dfrac{\partial^2 f(x)}{\partial x_1 \partial x_n} \\ \vdots & \ddots & \vdots \\ \dfrac{\partial^2 f(x)}{\partial x_n \partial x_1} & \cdots & \dfrac{\partial^2 f(x)}{\partial x_n \partial x_n} \end{bmatrix}$$

$H(x)$ is called the Hessian matrix or just the Hessian. This matrix is very useful in determining the local extreme (minimum or maximum) value of f. A necessary condition for $x = x_0$ being a local extreme of f is

$$m(x_0) = 0$$

If this condition holds, then

if $H(x_0)$ is positive definite, x_0 is a local minimum.

if $H(x_0)$ is negative definite, x_0 is a local maximum.

APPENDIX B: ESSENTIALS OF LARGE-SAMPLE THEORY[1]

The behavior of a statistic in small, or finite, samples and in large, or infinite, samples is not always the same. For example, in the classical linear regression model, we showed that the estimated error variance obtained by ordinary least squares is unbiased, whether in small or large samples, but that obtained by the method of maximum likelihood is biased in small samples but is unbiased as the sample size increases indefinitely.

In many cases, it is not easy to estimate the statistical properties of an estimator in finite samples, but it may be possible to do so in large, or infinite, samples. For example, we can derive the sampling distribution of the sample mean, $\bar{X}$, but it is not easy to derive the sampling distribution of $1/\bar{X}$, which is a nonlinear function of the sample mean.

In this appendix, we consider several aspects of **large-sample theory**, also known as **asymptotic theory**, the term *asymptotic* meaning the limiting behavior of a variable or a statistic of interest as the sample size increases indefinitely. In what follows, we discuss some salient aspects of asymptotic theory.

To develop this theory, we need some background, which is provided by some well-known inequalities and theorems.

B.1 Some Inequalities

B.1.1 Markov's Inequality

Let X be a nonnegative random variable, and suppose that $E(X)$ exists. Then for any $a > 0$,

$$\Pr(X > a) \le \frac{E(X)}{a} \qquad (B.1)$$

---

[1]See Amemiya, T. (1994). *Introduction to statistics and econometrics* (chap. 6). Cambridge, MA: Harvard University Press; Goldberger, A. S. (1991). *A course in econometrics* (chap. 9). Cambridge, MA: Harvard University Press.

Assuming X is a continuous random variable with density function $f(X)$,

$$
\begin{aligned}
E(X) &= \int_0^\infty xf(x)dx \\
&= \int_0^a xf(x)dx + \int_a^\infty xf(x)dx \\
&\geq \int_a^\infty xf(x)dx \\
&\geq \int_a^\infty af(x)dx \\
&\geq a\int_a^\infty f(x)dx \\
&= a\Pr(X \geq a) \\
\frac{E(X)}{a} &\geq \Pr(X \geq a)
\end{aligned}
$$

or

$$
\Pr(X > a) \leq \frac{E(X)}{a} \tag{B.2}
$$

which is the required result. Essentially, Markov's inequality states that the probability that X is much bigger than $E(X)$ is small.

A great advantage of Markov's inequality is that it enables us to place an upper bound on the probability that $g(X) \geq a$ as long as $g(X)$ is a nonnegative value and $Eg(X)$ exists.

B.1.2 Chebyshev's Inequality

As a corollary of Markov's inequality, we obtain Chebyshev's inequality. If X is a random variable with mean μ and variance σ^2, then for any value $k > 0$, it can be shown that

$$
\Pr(|X - \mu| \geq k\sigma) \leq \frac{1}{k^2}; \quad k > 0 \tag{B.3}^2
$$

This inequality is important because it provides a bound for all random variables. For example, it shows that the probability that the value of a random variable deviates more than three standard deviations ($k = 3$) from its mean value is less than 1/9.

---

[2]The inequality is also expressed as $\Pr(|X - \mu| \geq k) \leq \frac{\sigma^2}{k^2}; \quad k > 0.$

Proof: We can use Markov's inequality to prove Chebyshev's inequality. Letting $a = (k\sigma)^2$ and noting that $(X - \mu)^2$ is a nonnegative random variable, we have

$$[\Pr(X - \mu)^2 \geq k^2] \leq \frac{E[(X - \mu)^2]}{k^2} \tag{B.4}$$

Since $(X - \mu)^2 \geq k^2\sigma^2$ if and only if $|X - \mu| \geq k\sigma$, Equation (B.4) is equivalent to

$$\Pr\{|X - \mu| \geq k\sigma\} \leq \frac{E[(X - \mu)]^2}{k^2\sigma^2} = \frac{1}{k^2} \tag{B.5}$$

which is the required result.

The reason Markov's and Chebyshev's inequalities are important is because they enable us to establish bounds on probabilities just knowing only the mean or both the mean and variance of the probability distribution without knowing the actual probability distribution. If the latter were known, there would be no need to look for the bounds as the desired probabilities could be computed exactly.

B.1.3 Khinchine's Theorem (Weak Law of Large Numbers)

If $X_1, X_2, \ldots, X_n$ are iid random variables, with mean μ, then for any $\varepsilon > 0$, the sample mean $\bar{X}_n$ converges in probability to μ as $n \to \infty$.

That is,

$$\lim_{n\to\infty} \Pr|\bar{X}_n - \mu| > \varepsilon \to 0$$

Alternatively,

$$\lim_{n\to\infty} \Pr|\bar{X}_n - \mu| \leq \varepsilon = 1 \tag{B.6}$$

In short, $p\lim(\bar{X}_n) = \mu$, where $p\lim$ stands for probability limit.

Khinchine's theorem is also known as the **weak law of large numbers** (WLLN) to distinguish it from the **strong law of large numbers** (SLLN), which we will discuss shortly.

The WLLN is easy to prove. We know that $E(\bar{X}) = E[(X_1 + X_2 + \cdots + X_n) / n] = n\mu / n = \mu$ and $\text{var}(\bar{X}) = \sigma^2 / n$. Therefore,

$$\lim_{n\to\infty} \frac{\sigma^2}{n} \to 0$$

We can establish this theorem, using Chebyshev's inequality:

$$\Pr[|\bar{X} - \mu| > \varepsilon] \le \frac{\sigma^2}{n\varepsilon^2}; \quad \varepsilon > 0 \tag{B.7}$$

B.1.4 Strong Law of Large Numbers

The WLLN is based on the concept of the *convergence in probability* limit, whereas the SLLN is based on the concept of almost sure convergence. These concepts of convergence are discussed below.

B.1.5 Jensen's Inequality

If $Y = g(X)$ is a concave function so that it lies everywhere below its tangent line and $E(X) = \mu$, then

$$E(Y) \le g(\mu) \tag{B.8}$$

It is well-known that the logarithmic function is concave, and so

$$E[\log(X)] \le \log[E(X)] \tag{B.9}$$

Similarly, if $Y = g(X)$ is a convex function so that it lies everywhere above its tangent line, then

$$E(Y) \ge g(\mu) \tag{B.10}$$

Thus, the square function is convex. So we have

$$E(X^2) \ge [E(X)]^2 \tag{B.11}$$

Both (B.9) and (B.10) hold regardless of the distribution of X.

B.1.6 Cauchy–Schwarz Inequality

If the random variables X and Y have finite variances, then

$$[E(XY)]^2 \le E(X^2)E(Y^2) \tag{B.12}$$

with equality if and only if $\Pr(aX + bY) = 1$ for some real constants a and b, at least one of which is nonzero. This inequality is used to establish that the correlation coefficient between two random variables lies between -1 and $+1$.

Without loss of generality, suppose $E(X) = E(Y) = 0$.

In this case,

$$E(XY) = \text{cov}(X,Y) = \sigma_{XY}; \quad E(X^2) = \text{var}(X); \quad E(Y^2) = \text{var}(Y)$$

and (B.12) reduces to

$$0 \le \frac{\sigma_{XY}^2}{\sigma_X^2 \sigma_Y^2} \le 1$$

$$0 \le \rho^2 \le 1$$

where ρ is the population coefficient of correlation defined as

$$\rho = \frac{\sigma_{xy}}{\sigma_x \sigma_y}$$

B.1.7 The Central Limit Theorem

One of the most remarkable results in probability theory is the **central limit theorem (CLT)**, proposed by the French mathematician Laplace. In simple terms, the CLT asserts that the sum of a large number of independent variables has a (probability) distribution that is approximately normal.

Formally, let $X_1, X_2, \ldots, X_n$ be a sequence of iid random variables each with mean μ and variance σ^2. Let $\bar{X}_n = \Sigma_{i=1}^n X_i / n$, which is the sample mean of the random variable X. Then as n increases indefinitely (i.e., $n \to \infty$),

$$\bar{X}_n \approx N\left(\mu, \frac{\sigma^2}{n}\right) \tag{B.13}$$

where $\approx$ means approximately. Note that this result holds true regardless of the form of the probability distribution function.

As a result, the following variable, Z, follows the standard normal distribution, namely,

$$Z = \frac{\bar{X}_n - \mu}{\sigma / \sqrt{n}} \approx N(0,1) \tag{B.14}$$

That is, Z follows the standard normal distribution. Notice that Equations (B.13), (B.14), and the following equations are all equivalent:

$$(\bar{X} - \mu) \approx N(0, \sigma^2/n)$$

$$\sqrt{n}(\bar{X}_n - \mu) \approx N(0, \sigma^2) \tag{B.15}$$

$$\frac{\sqrt{n}(\bar{X}_n - \mu)}{\sigma} \approx N(0,1)$$

We will not provide a proof of this theorem as it involves *moment-generating functions* or *characteristic functions*.[3]

Now we turn to a discussion of convergence. This topic is often discussed under the title of **limit theorems**. The general idea behind the limit theorems is to study the behavior of a stochastic quantity as the sample size n increases indefinitely.

B.2 Types of Convergence

B.2.1 Convergence of a Sequence of Real Numbers

A sequence of real numbers $\{a_n\}$, $n = 1, 2, 3, \ldots$, converges to a real number a if for any $\varepsilon > 0$, there exists an integer N such that for all $n > N$, we have

$$|a_n - a| < \varepsilon \tag{B.16}$$

which means as $n \to \infty$, $\lim_{n \to \infty} a_n = a$.

As an example, let $a_t = 6 + (0.4)^t$, as $t \to \infty$, $a_t \to 6$. Hence, $\lim_{t \to \infty} a_t = 6$. For another example, $\lim_{n \to \infty} \left[1 + \frac{1}{n} \right] = 1$. Consider now $a_t = (-0.999)^t$. As t tends to infinity, this sequence tends to 0, but in an oscillating manner.

B.2.2 Strong or Almost Sure Convergence

If for all $\varepsilon > 0$, then $\Pr(\lim_{n \to \infty} X_n - X) = 1$, X_n converges to X *almost surely* or with probability 1 to X, written as $X_n \xrightarrow{as} X$, where "as" = almost surely. Almost sure convergence is used in advanced analysis, but convergence in probability, discussed below, is easier to work with and is sufficient for most purposes.

B.2.3 Convergence in Probability
of a Sequence of Random Variables

Consider a sequence of random variables $X_1, X_2, \ldots, X_n$ with (cumulative) distribution functions $F_1(\ ), F_2(\ ), \ldots, F_n(\ )$, the index n indicating the number of terms in the sequence.

This sequence is said to converge in probability to a constant c, if

$$\lim_{n \to \infty} \Pr[|X_n - c| > \varepsilon] = 0 \text{ for any } \varepsilon > o | \tag{B.17}$$

---

[3]See Hogg, R. V., Mckean, J., & Craig, A. T. (2014). *Introduction to mathematical statistics* (7th ed.). Noida, India: Pearson Education.

What (B.17) states is that the sequence of random variables $X_1, X_2, ..., X_n$ converges in probability to the constant c if the limit of this sequence of probabilities is zero for any positive value of ε.

To account for the probabilistic nature of random variables, we can modify the definition given in (B.17) as follows: A sequence of random variables $\{X_n\}, n = 1, 2, 3, ...$ is said to converge to a random variable X in *probability*, denoted as $X_n \overset{P}{\to} X$ if for any $\varepsilon > 0$, as $n \to \infty$

$$\Pr[|X_n - X| \geq \varepsilon] \to 0 \tag{B.18}$$

or equivalently, for any $\varepsilon > 0$, as $n \to \infty$

$$\Pr[|X_n - X| < \varepsilon] \to 1 \tag{B.19}$$

In simple terms, this means that the difference between X_n and X is likely to be small as the sample size n increases indefinitely. This concept plays a key role in determining the consistency of an estimator and the various laws of large numbers.

Equations (B.18) and (B.19) are often expressed as

$$\underset{n \to \infty}{plim} X_n = X \tag{B.20}$$

where $plim$ means probability limit.

Note that $plim(\ )$ is an operator, like $E(\ )$, the expectations operator. But the $plim$ has some properties that are not shared by the expectations operator, which makes it easy to use $plim$ in situations where it is difficult or impossible to obtain results based on the expectations operator, as the following discussion will show. One reason for this is that the $E(\ )$ is a **linear operator**.

The $plim$ has the following properties, often attributed to Slutsky, a Russian mathematician. If $X_n \overset{P}{\to} X$ and $Y_n \overset{P}{\to} Y$, then

1. $aX_n \overset{P}{\to} aX$ (a, real number).
2. If $X_n \overset{P}{\to} X$ and $Y_n \overset{P}{\to} Y$, then $X_n + Y_n \overset{P}{\to} X + Y$.
3. If $X_n \overset{P}{\to} X$ and $Y_n \overset{P}{\to} Y$, then $X_n Y_n \overset{P}{\to} XY$.
4. If $X_n \overset{P}{\to} X$ and $Y_n \overset{P}{\to} Y$, then $\dfrac{X_n}{Y_n} \overset{P}{\to} \dfrac{X}{Y}$, if Y is not 0.
5. If $X_n \overset{P}{\to} X$, then $g(X_n) \overset{P}{\to} g(X)$.

6. If $h(\ )$ is a continuous function, then $X_n \to X$ implies that $h(X_n) \to h(X)$, that is, convergence in probability is preserved under continuous transformation. Notice that properties (3) and (4) do not hold for the expectations operator. Thus, $E(XY)$ is not equal to $E(X) \cdot E(Y)$, unless X and Y are independent. Similarly, $E(X/Y)$ is *not* equal to $E(X)/E(Y)$.

Besides convergence in probability, there are two other modes of convergence: *convergence in mean square* and *convergence in distribution*.

B.2.4 Convergence in Mean Square

A sequence $\{X_n\}$, $n = 1, 2, 3, \ldots$ is said to converge to X in *mean square* if

$$\lim_{n \to \infty} E(X_n - X)^2 \to 0 \tag{B.21}$$

which is often expressed as $X_n \overset{M}{\to} X$. For example, the sample mean $\overline{X}_n$ from a population with mean μ and variance σ^2 converges in mean square to μ because $E[(\overline{X}_n - \mu)^2] = \mathrm{var}(\overline{X}_n) = \sigma^2/n \to 0$ as $n \to \infty$.

B.2.5 Convergence in Distribution

A sequence $\{X_n\}$, $n = 1, 2, 3, \ldots$ is said to converge to X if the distribution function F_n of X_n converges to the distribution function F of X at every continuity point of F. We express this as

$$X_n \overset{D}{\to} X \tag{B.22}$$

F is called the limit distribution of $\{X_n\}$.

For example, if $X \sim N(0,1)$, that is, it follows the standard normal distribution, we can write (B.22) as $X_n \overset{D}{\to} N(0,1)$.

It may be noted that

$$\text{If } X_n \overset{D}{\to} X \text{ and } Y_n \overset{D}{\to} c, \text{ then } X_n + Y_n \overset{D}{\to} X + c$$

$$\text{If } X_n \overset{D}{\to} X \text{ and } Y_n \overset{D}{\to} c, \text{ then } X_n Y_n \overset{D}{\to} cX$$

$$\text{If } X_n \overset{D}{\to} X, \text{ then } g(X_n) \overset{D}{\to} g(X)$$

where c is a constant.

B.2.6 Relationships Among Three Modes of Convergence

There is a hierarchy among the various modes of convergence: **almost sure convergence $\to$ convergence in mean square $\to$ convergence in**

probability → **convergence in distribution**, the arrow indicating which convergence dominates. This chain of convergence is not reversible: for example, convergence in distribution does not imply convergence in probability, and so on.[4] The various convergence theorems can be proved with the aid of Chebyshev's inequality and Khinchine's theorem. We will not prove this explicitly, for some of the proofs are involved and can be found in the references.

The concepts of convergence help us in distinguishing between the WLLN and the SLLN: the WLLN converges to μ in probability and the SLLN says that this is also true almost surely.

B.3 The Order of Magnitude of a Sequence

In studying the rate of convergence of sequences of variables, it is useful to know the concept of the **order of magnitude** of such sequences. We first consider the case of a nonstochastic sequence of real numbers a_n, as in the following examples.

Consider the sequence $a_n = 6 + n - 4n^2$. As n becomes increasingly larger, the terms 6 and n become small (in absolute value) compared with $-4n^2$. We call the last term the *leading term* of the sequence, and it determines the order of magnitude of the sequence.

In general, we say that the sequence a_n is *at most* of order n^k, written as $O(n^k)$, or big O n^k, if the sequence $n^{-k}a_n$ is bounded. For the above sequence, if we take $k = 2$, we obtain

$$n^{-2}(6 + n + 4n^2) = \frac{6}{n^2} + \frac{1}{n} - 4$$

Therefore, as $n \to \infty$, this sequence converges to -4, so that it is bounded. Hence, this sequence is $O(n^2)$. As you can see, it is the leading term that determines the order of magnitude.

A related concept is *of smaller order than* n^k, written as $o(n^k)$, or small o (n^k), which means that the sequence $n^{-k}a_n$ converges to zero. In our example, $a_n = o(n^3)$ because

$$n^{-3}(6 + n - 4n^2) = \frac{6}{n^3} + \frac{1}{n^2} - \frac{4}{n}$$

As $n \to \infty$, this sequence tends to zero. This is true even if $k = 2.5$ or $k = -8$.

---

[4]For proof, see Ramanathan, R. (1993). *Statistical methods in econometrics* (chap. 7). New York, NY: Academic Press.

In the special case of $k = 0$, $a_n = O(n^0) = O(1)$, that is, the sequence is bounded, and in the case of small o, $a_n = o(n^0) = o(1)$, the sequence has a zero limit.

The algebra of sequences is similar to ordinary algebra. Thus, if $a_n = 4n$ and $b_n = 2 + n^{-1}$, then $a_n + b_n = 4n + 2 + n^{-1}$ and $a_n b_n = 8n + 4$.

In general, if $a_n = O(n^h)$ and $b_n = O(n^k)$, then $a_n b_n = O(n^{h+k})$ and

$$a_n + b_n = O(n^l)$$

where $l = \max(h, k)$.

B.4 The Order of Magnitude of a Stochastic Sequence

The ideas of order of magnitude for a nonstochastic sequence can be carried over to stochastic sequences.

Let X_n, $n = 1, 2, \ldots, N$ denote a sequence of real-valued random variables. Then X_n is said to be bounded in probability if for every $\varepsilon > 0$, there exists a positive constant c and a positive integer N such that

$$\Pr[|X_n| > c] \le \varepsilon \tag{B.23}$$

for all $n \ge N$ and where ε is an arbitrarily small positive number.

In words, X_n is bounded in probability if for any arbitrarily small positive ε we can always find a positive constant c such that the probability of the absolute value of X_n being larger than c is less than ε.

Equation (B.23) can be equivalently expressed as

$$\Pr[|X_n| \le c] > 1 - \varepsilon \tag{B.24}$$

for all $n \ge N$.

Notice the following statements:

1. $X_n = O_p(1)$ means that X_p is bounded in probability.

2. $X_n = o_p(1)$ means $X_n \xrightarrow{p} 0$, that is, $\Pr[|X_n| > \varepsilon] \to 0$ as $n \to \infty$.

APPENDIX C: SMALL- AND LARGE-SAMPLE PROPERTIES OF ESTIMATORS

An estimator can be judged by several properties. These properties fall into two categories: (1) small sample and (2) large sample.

C.1 Small-Sample Properties of Estimators

C.1.1 Unbiasedness

An estimator $\hat{\theta}$ is said to be an unbiased estimator of θ if the expected value of $\hat{\theta}$ is equal to θ, that is,

$$E(\hat{\theta}) = \theta \tag{C.1}$$

If this equality does not hold, then the estimator is said to be biased, which can be expressed as

$$\text{bias}(\hat{\theta}) = E(\hat{\theta}) - \theta \tag{C.2}$$

It is important to note that unbiasedness is a property of repeated sampling, not of any given sample: Keeping the sample size fixed, we draw several samples from a given population, each time obtaining an estimate of the unknown parameter. For the estimator to be unbiased, the average value of these estimates must be equal to the true value.

C.1.2 Minimum Variance

$\hat{\theta}_1$ is said to be a minimum variance estimator of θ if its variance is smaller than or at most equal to the variance of $\hat{\theta}_2$, which is another estimator of θ. A minimum variance estimator is not necessarily unbiased.

C.1.3 Best Unbiased or Efficient Estimator

If $\hat{\theta}_1$ and $\hat{\theta}_2$ are two *unbiased* estimators of θ, and the variance of $\hat{\theta}_1$ is smaller than the variance of $\hat{\theta}_2$, then $\hat{\theta}_1$ is a **minimum variance unbiased**, or **best unbiased**, or **efficient**, estimator.

C.1.4 Linearity

An estimator $\hat{\theta}$ is said to be a linear estimator of θ if it is a linear function of the sample observations. For example, the sample mean defined as

$$\bar{X} = \frac{1}{n}\sum_1^n X_i = \frac{1}{n}(X_1 + X_2 + \cdots + X_n)$$

is a linear estimator because it is a linear function of the X values.

C.1.5 Best Linear Unbiased Estimator

If $\hat{\theta}$ is linear, is unbiased, and has minimum variance in the class of all linear unbiased estimators of θ, then it is called a best linear unbiased estimator, or BLUE for short.

C.1.6 Minimum Mean Square Error Estimator

The minimum mean square error (MSE) of an estimator $\hat{\theta}$ is defined as

$$\text{MSE}(\hat{\theta}) = E(\hat{\theta} - \theta)^2 \tag{C.3}$$

whereas the variance of $\hat{\theta}$ is defined as

$$\text{var}(\hat{\theta}) = E[\hat{\theta} - E(\hat{\theta})]^2 \tag{C.4}$$

The difference between the two is that the variance measures the dispersion of the distribution of an estimator around its mean or expected value, whereas MSE measures it around the true value of the parameter.

Simple algebraic manipulation will show that

$$\text{MSE}(\hat{\theta}) = E[\hat{\theta} - E(\hat{\theta})]^2 + [E(\hat{\theta}) - \theta]^2 \tag{C.5}$$
$$= \text{var}(\hat{\theta}) + [\text{bias } (\hat{\theta})]^2$$

Of course, if the bias is zero, MSE and variance of an estimator are the same. The MSE thus takes into account both bias and variance. Thus, MSE provides a trade-off between the variance and the bias of an estimator. In practice, the MSE criterion is used if the best unbiased criterion is incapable of producing estimators with smaller variances. In discussing multicollinearity in Chapter 7, we stated that estimators based on ridge regression are biased but they tend to have a smaller MSE than the OLS estimators.

C.1.7 Efficient Estimator

An estimator $\hat{\theta}_1$ is an efficient estimator of θ if

$$E(\hat{\theta}_1) = \theta \text{ and var}(\hat{\theta}_1) \leq \text{var } (\hat{\theta}_2) \tag{C.6}$$

where $\hat{\theta}_2$ is any other unbiased estimator of θ.

An efficient estimator defined this way is also known as a *minimum variance unbiased estimator* (MVUE) or *best unbiased estimator.*

How does one find such an estimator? Here, we can get some guidance from the well-known Cramer–Rao (CR) theorem, which provides *a sufficient but not necessary condition* for an unbiased estimator to be efficient.[1]

---

[1]The proof of the CR theorem is rather involved and can be found in Theil, H. (1971). *Principles of econometrics* (pp. 384–386). New York, NY: John Wiley.

To understand the idea behind the CR theorem, consider a random variable X with density function $f(X,\theta)$, where θ is the unknown parameter; for simplicity, we assume that there is only one unknown parameter, but this can be generalized for a density function with several unknown parameters. Let $X_1, X_2, \ldots X_n$ denote a random sample drawn from this density function, n being the sample size.

Let $L(X_1, X_2, \ldots, X_n \mid \theta)$ be the likelihood function of the sample, and further assume that it is twice differentiable and it satisfies certain **regularity conditions**, such as differentiation under the integral sign, the limit of integration is independent of the true θ, and the order of integration and differentiation can be interchanged. Let $\hat{\theta}$ be an unbiased estimator of θ. Then the variance of $\hat{\theta}$ must satisfy the inequality

$$\text{var}(\hat{\theta}) \geq -\cfrac{1}{E\left(\cfrac{\partial^2 l}{\partial \theta^2}\right)} \tag{C.7}$$

where $l = \ln L$, that is, the (natural) log of the likelihood function, L. The right-hand side of this equation is known as the Cramer–Rao lower bound (CRLB).

The quantity $I = E\left(\dfrac{\partial^2 \ln l}{\partial \theta^2}\right)$ is called **Fisher's information matrix**[2]—the amount of information that a sample provides about the value of an unknown parameter(s). Thus, the greater the variance, the less the information.

The basis of the CR inequality is Fisher's information matrix, although since 1948 it has been called the CR inequality because Cramer and Rao proved this inequality independently. An estimator that achieves the CRLB is called an *efficient estimator*.

The CR inequality can be generalized to account for more than one unknown parameter, say, $\theta_1, \theta_2, \ldots, \theta_p$, in which case the information matrix can be expressed as

$$I_{jk} = E\left\{\left(\frac{\partial \ln l}{\partial \theta_j}\right)\left(\frac{\partial \ln l}{\partial \theta_k}\right)\right\}' \tag{C.8}$$

Commenting on the CR inequality, Theil writes as follows:

As soon as we have found an unbiased estimator whose variance is equal to minus the reciprocal of the expected second-order derivative of the log-likelihood of the sample, we know that we cannot find an

---

[2]Sir Ronald Fisher, a British statistician, developed this matrix in 1922.

unbiased estimator with a smaller variance. Hence, no further improvements are possible if we confine ourselves to unbiased estimators and if minimum sample variance is our goal.[3]

However, an efficient estimator in this sense may not exist. This is because the CRLB is only a sufficient condition and not a necessary one because in some situations the lower bound on the variance cannot be attained by any unbiased estimator. Besides, one or more regularity conditions underlying the CR theorem may not be satisfied. However, in some cases, we can actually find the variance of an estimator that satisfies the CR inequality, as the following example shows.

Consider the sample mean $\bar{X}$ from a normal population with mean μ. We want to prove that $\bar{X}$ is an MVUE of μ.

Proof: Since

$$f(X) = \frac{1}{\sigma\sqrt{2\pi}} e^{-\frac{1}{2}\left(\frac{X-\mu}{\sigma}\right)^2} \qquad -\infty < X < \infty$$

it follows that

$$\ln f(X) = -\ln \sigma\sqrt{2\pi} - \frac{1}{2}\left(\frac{X-\mu}{\sigma}\right)^2$$

Now

$$\frac{\partial \ln f(X)}{\partial \mu} - \frac{1}{\sigma}\left(\frac{X-\mu}{\sigma}\right)$$

Hence,

$$E\left[\left(\frac{\partial \ln f(X)}{\partial \mu}\right)^2\right] = \frac{1}{\sigma^2} \cdot E\left[\left(\frac{X-\mu}{\sigma}\right)^2\right] = \frac{1}{\sigma^2} \cdot 1 = \frac{1}{\sigma^2}$$

Therefore,

$$\frac{1}{n \cdot E\left[\left(\frac{\partial \ln f(X)}{\partial \mu}\right)^2\right]} = \frac{1}{n \cdot \frac{1}{\sigma^2}} = \frac{\sigma^2}{n} \qquad (C.9)$$

[3]See Theil, H. (1971). *Principles of econometrics* (pp. 386–387). New York, NY: John Wiley.

Since $\bar{X}$ is unbiased and $\mathrm{var}(\bar{X}) = \dfrac{\sigma^2}{n}$, according to the CR theorem, $\bar{X}$ is an MVUE of μ.

We have shown earlier that in the classical normal linear regression, the ML estimators of the regression parameters attain the CRLB and hence they are most efficient.

C.1.8 Sufficient Estimator(s)

Consider a random variable X with the density function $f(X,\theta)$, where θ is a single parameter of this density function; for simplicity of exposition, we assume that there is just a single parameter, although we can extend it to multiple parameters. Suppose from this density function, we draw a random sample of n observations, $X_1, X_2, \ldots X_n$ and compute a sample statistic, say $\hat{\theta}$. We say that $\hat{\theta}$ is a *sufficient* statistic for θ if it *encapsulates* all the information about θ, that is, it condenses the sample data in such a way that no information about θ is lost. If such a statistic exists, there is no need to look for examining the entire sample or another statistic based on this sample.

Formally, let $X = \{x_1, x_2, \ldots, x_n\}$ be a random sample from a population with density function $f(X,\theta)$. Then the statistic or estimator $\hat{\theta}$ is a sufficient estimator of θ if and only if the joint density or probability distribution of the random sample can be factored so that

$$L(\theta; X) = f(X_1, X_2, \ldots, X_n; \theta) = h(\theta, \hat{\theta}) \cdot g(X) \qquad (C.10)$$

where $g(X)$ stands for $g(X_1, X_2, \ldots, X_n)$ and where $h(\theta, \hat{\theta})$ depends only on $\hat{\theta}$ and θ and $g(X)$ does not depend on θ.

Equation (C.10) is known as the **Fisher–Neyman factorization theorem**. This theorem states that a statistic or estimator $\hat{\theta}$ is sufficient for θ if and only if the joint density of the sample can be factored into two components—one depends only on the estimator and the true parameter and the other is independent of the parameter.

Example: Consider a sample of n observations from a normal population with mean μ and *known* variance σ^2. It can be shown that $\bar{X}$, the sample mean, is a sufficient estimator of μ.

Since

$$L(X_1, X_2, \ldots, X_n) = \left(\frac{1}{\sigma\sqrt{2\pi}}\right)^n \exp\left\{\left(-\frac{1}{2}\right)\Sigma\left(\frac{X_i - \mu}{\sigma}\right)^2\right\} \qquad (C.11)$$

the log-likelihood function becomes

$$l(X_1, X_2, \ldots, X_n) = n \ln\left(\frac{1}{\sigma\sqrt{2\pi}}\right) - \frac{1}{2}\Sigma\left(\frac{X_i - \mu}{\sigma}\right)^2 \tag{C.12}$$

With simple algebraic manipulation, it can be shown that

$$\Sigma(X_i - \mu)^2 = \Sigma(X_i - \bar{X})^2 + n(\bar{X} - \mu)^2 \tag{C.13}$$

As a result, we can express Equation (C.12) as

$$l(X_1, X_2, \ldots, X_n) = \ln\left(\frac{\sqrt{n}}{\sigma\sqrt{2\pi}}\right) - \frac{1}{2}\left(\frac{\bar{X} - \mu}{\sigma\sqrt{n}}\right)^2$$
$$+ \left\{ \ln\frac{1}{\sqrt{n}}\left(\frac{1}{\sigma\sqrt{2\pi}}\right)^{n-1} - \frac{1}{2}\Sigma\left(\frac{X_i - \bar{X}}{\sigma}\right)^2 \right\} \tag{C.14}$$

From Equation (C.14), we see that the first factor on the right-hand side involves $\bar{X}$ and the population mean μ, and the second factor does not involve μ. By the Fisher–Neyman factorization theorem, we can therefore say that $\bar{X}$ is a sufficient estimator of μ of a normal population with the known variance σ^2.

C.1.8.1 Properties of Sufficient Estimators

It is important to know some of the propertics of sufficient statistics.[4]

1. If $\hat{\theta}$ is sufficient for θ, then the likelihood function for θ based on the distribution of $\hat{\theta}$ is proportional to $L(\theta; X)$, where $L(\theta; X)$ is given by (C.10).

This result makes sense, for $\hat{\theta}$ carries all the sample information about θ. In other words, we get the same information about θ from $\hat{\theta}$ as we would get from the entire sample of y.

2. If $\hat{\theta}$ is sufficient for θ, the conditional distribution of outcomes y, given the observed value of $\hat{\theta}$, does not depend on θ. A sufficient statistic is often defined this way.

---

[4]For details, see Kalbfleisch, J. G. (1985). *Probability and statistical inference, vol. 2: Statistical inference* (2nd ed., pp. 285–289). New York, NY: Springer-Verlag.

3. Every single-valued function of the sufficient statistic is also a sufficient statistic of the true value of the estimator. Stated differently, a one-to-one transformation of sufficient statistics produces another set of sufficient statistics. For example, if $X_1, X_2, ..., X_n$ are independent Bernoulli random variables with the parameter θ, then

$\hat{\theta} = \dfrac{\sum_1^n X_i}{n}$ is a sufficient estimator of θ and $X = X_1 + X_2 + \cdots + X_n$ is

also a sufficient estimator of the binomial mean $\mu = n\theta$.

4. As Kalbfleisch notes, the ML estimator $\hat{\theta}$ is part of any set of sufficient statistics $\hat{\theta}_1, \hat{\theta}_2, \ldots, \hat{\theta}_k$, in that its value can be computed from just the $\hat{\theta}_i$s. This is because the $\hat{\theta}_i$s determine $L(\theta, y)$ up to a proportionality constant, and $\hat{\theta}$ does not depend on this constant.

C.1.9 Uniformly Minimum Variance Unbiased Estimator

An unbiased estimator that has minimum variance in the entire class of unbiased estimators, whether linear or nonlinear, is called a uniformly minimum variance unbiased estimator (UMVUE). More technically, an unbiased estimator that has variance equal to CRLB must have minimum variance among all unbiased estimators. Sufficiency is a powerful property in finding UMVUE estimators.

C.2 Large-Sample Properties of Estimators

An estimator may not satisfy one or more of the desirable statistical properties in small samples, but as the sample size increases indefinitely, the estimator may possess several desirable statistical properties. These properties are known as **large-sample**, or **asymptotic**, **properties**, such as the following:

C.2.1 Asymptotic Unbiasedness

An estimator $\hat{\theta}$ is said to be an asymptotically unbiased estimator of θ if

$$\lim_{n \to \infty} E(\hat{\theta}_n) = \theta \tag{C.15}$$

where the subscript n on the estimator means that the estimator is based on a sample size n and where "lim" means limit and $n \to \infty$ means that n increases indefinitely. In words, $\hat{\theta}$ is an asymptotically unbiased estimator of θ if its expected, or mean, value approaches the true value as the sample size gets larger and larger.

As an example, consider the following measure of the sample variance of a random variable X with mean of μ and variance σ^2:

$$S^2 = \frac{\Sigma(X_i - \bar{X})^2}{n} \tag{C.16}$$

Algebraic manipulation will show that

$$\Sigma(X_i - \bar{X}^2)^2 = \Sigma[(X_i - \mu) - (\bar{X} - \mu)]^2$$
$$= \Sigma(X_i - \mu)^2 - n(\bar{X} - \mu)^2$$

Therefore,

$$S^2 = \frac{\Sigma(X_i - \mu)^2 - n(\bar{X} - \mu)^2}{n}$$
$$= \frac{E\Sigma(X_i - \mu)^2 - nE(\bar{X} - \mu)^2}{n}$$
$$E(S^2) = \sigma^2 - \frac{\sigma^2}{n} \tag{C.17}$$
$$= \sigma^2\left(1 - \frac{1}{n}\right)$$

which shows that the sample variance as defined in (C.16) is a biased estimator of σ^2. *Note:* $E(X_i - \mu)^2 = \sigma^2$ and $E(\bar{X} - \mu)^2 = \sigma^2/n$, which is the variance of the sample mean.

However,

$$E(S^2) = \sigma^2 \tag{C.18}$$
$$\scriptstyle n \to \infty$$

Obviously, S^2 is biased, but as n increases indefinitely, $E(S^2)$ approaches the true σ^2, hence it is asymptotically unbiased. We have showed earlier that the ML estimator of the error variance of the normal LRM is biased, but the bias gets smaller and smaller as the sample size increases indefinitely.

It may be noted that if we define the sample variance as

$$s^2 = \frac{\Sigma(X_i - \bar{X})^2}{n-1} \tag{C.19}$$

then it can be shown that this estimator is unbiased, as $E(s^2) = \sigma^2$, regardless of the sample size. Note that we have used s^2 to distinguish it from the

S^2 defined in (C.16); the difference between the two lies in that (C.19) takes into account the degrees of freedom.

$$(n-1)s^2 = \Sigma[X_i - \mu - (\overline{X} - \mu)]^2$$
$$= \Sigma[(X_i - \mu)^2 - 2(X_i - \mu)(\overline{X} - \mu) + (\overline{X} - \mu)^2]$$
$$= \Sigma(X_i - \mu)^2 - n(\overline{X} - \mu)^2$$

since

$$2\Sigma(X_i - \mu)(\overline{X} - \mu) = 2n(\overline{X} - \mu)^2$$

Now,

$$(n-1)Es^2 = E\Sigma(X_i - \mu)^2 - nE(\overline{X} - \mu)^2$$
$$= n\sigma^2 - n(\sigma^2/n)$$
$$= n\sigma^2 - \sigma^2$$
$$= \sigma^2(n-1)$$

Therefore,

$$Es^2 = \sigma^2 \qquad (C.20)$$

In deriving this result, we make use of an important relationship that for any random variable, say, V,

$$E(V^2) = \text{var}(V) + (E[V])^2 \qquad (C.21)$$

C.2.2 Consistency

The property of consistency is concerned with the asymptotic (i.e., as $n \to \infty$) accuracy of an estimator, that is, whether it converges to the parameter that it is estimating.

$\hat{\theta}$ is said to be a consistent estimator of the population parameter θ if it approaches the true value θ as the sample size gets larger and larger. More formally, an estimator $\hat{\theta}$ is a consistent estimator of θ if the probability that the absolute value of the difference between the two is less than δ (an arbitrarily small positive quantity) and approaches unity. Symbolically,

$$\lim_{n \to \infty} \Pr\{|\hat{\theta} - \theta| < \delta\} = 1 \quad \delta > 0 \qquad (C.22)$$

where Pr stands for probability. This is often expressed as

$$p\lim_{n \to \infty} \hat{\theta} = \theta \qquad \text{(C.23)}$$

where $p$lim means probability limit.

It is important to note that the properties of unbiasedness and consistency are conceptually quite different. The property of unbiasedness can hold for any sample size, whereas consistency is strictly a large-sample property.

A *sufficient condition* for consistency is that the bias and variance both tend to zero as the sample size increases indefinitely. Technically,

$$\lim_{n \to \infty} E(\hat{\theta}) = \theta \text{ and } \lim_{n \to \infty} \text{var}(\hat{\theta}_n) = 0 \qquad \text{(C.24)}$$

Alternatively, a sufficient condition for consistency is that the $\text{MSE}(\hat{\theta})$ tends to zero as n increases indefinitely.

Example: Let $X_1, X_2, \ldots, X_n$ be a random sample from a distribution with mean μ and variance σ^2. It is easy to show that the sample mean $\bar{X}$ is a consistent estimator of μ. Since $E(\bar{X}) = \mu$ regardless of the sample size, it is unbiased and $\text{var}(\bar{X}) = \sigma^2/n$ regardless of the sample size. Furthermore, as n increases indefinitely, $\text{var}(\bar{X})$ tends toward zero. Hence, the sample mean is a consistent estimator of the population mean. In short, $p\lim(\bar{X}) = \mu$.

C.2.2.1 Probability Limit (plim)

In establishing the consistency property of estimators, the following properties of the probability limit ($p$lim) are noteworthy.

1. *Invariance* (Slutsky property). If $\hat{\theta}$ is a consistent estimator of θ and if $h(\hat{\theta})$ is any continuous function of $\hat{\theta}$, then

$$p\lim_{n \to \infty} h(\hat{\theta}) = h(\theta) \qquad \text{(C.25)}$$

What this means is that if $\hat{\theta}$ is a consistent estimator of θ, then $1/\hat{\theta}$ is also a consistent estimator of $1/\theta$ and that $\log(\hat{\theta})$ is a consistent estimator of $\log(\theta)$

The invariance property does not hold true of the expectations operator E. Thus, if $\hat{\theta}$ is an unbiased estimator of θ, it is not the case that $1/\hat{\theta}$ is an unbiased estimator of $1/\theta$, that is

$$E\left(\frac{1}{\hat{\theta}}\right) \neq \frac{1}{E(\hat{\theta})} \neq \frac{1}{\theta} \qquad \text{(C.26)}$$

2. If b is a constant, then

$$\plim_{n \to \infty} = b \qquad (C.27)$$

That is, the probability limit of a constant is the same constant.

3. If $\hat{\theta}_1$ and $\hat{\theta}_2$ are consistent estimators, then

$$\plim(\hat{\theta}_1 + \hat{\theta}_2) = \plim\hat{\theta}_1 + \plim\hat{\theta}_2$$
$$\plim(\hat{\theta}_1 - \hat{\theta}_2) = \plim\hat{\theta}_1 - \plim\hat{\theta}_2$$
$$\plim(\hat{\theta}_1\hat{\theta}_2) = \plim\hat{\theta}_1\plim\hat{\theta}_2 \qquad (C.28)$$
$$\plim\left(\frac{\hat{\theta}_1}{\hat{\theta}_2}\right) = \frac{\plim\hat{\theta}_1}{\plim\hat{\theta}_2}; \quad \text{for } \plim\hat{\theta}_2 \neq 0$$

Again, note that the last two properties do not hold true of the expectations operator E. Thus,

$$E\left(\frac{\hat{\theta}_1}{\hat{\theta}_2}\right) \neq \frac{E(\hat{\theta}_1)}{E(\hat{\theta}_2)} \qquad (C.29)$$
$$E(\hat{\theta}_1 \hat{\theta}_2) \neq E(\hat{\theta}_1) E(\hat{\theta}_2)$$

If, however, $\hat{\theta}_1$ and $\hat{\theta}_2$ are independently distributed,

$$E(\hat{\theta}_1 \hat{\theta}_2) = E(\hat{\theta}_1) E(\hat{\theta}_2) \qquad (C.30)$$

Given the classical setup, we have shown that the OLS estimators are unbiased. We can also show that these estimators are consistent. To illustrate, consider the following simplest linear regression:

$$Y_i = B_1 + B_2 X_i + u_i$$

Under classical assumptions, it is easy to show that

$$b_1 = \bar{Y} - b_2 \bar{X}$$
$$b_2 = \frac{\Sigma y_i x_i}{\Sigma x_i^2}$$

where

$$y_i = (Y_i - \bar{Y}); \quad x_i = (X_i - \bar{X})$$
$$\text{var}(b_2) = \frac{\sigma^2}{\Sigma x_i^2}$$

To prove that b_2 is a consistent estimator of B_2, we need to show that the variance of b_2 tends to zero as n, the number of sample observations, increases indefinitely. $\Sigma x_i^2 / n \neq 0$, because the variance of X is bounded. We proceed as follows:

$$\text{var}(b_2) = \frac{\sigma^2}{\Sigma x_i^2} = \frac{\sigma^2 / n}{\Sigma x_i^2 / n}$$

Dividing the numerator and the denominator by n, we do not change the equality.

Now

$$\lim_{x \to \infty} \text{var}(b_2) = \lim_{x \to \infty} \left(\frac{\sigma^2 / n}{\Sigma x_i^2 / n} \right) = \frac{\lim(\sigma^2 / n)}{\lim(\Sigma x_i^2 / n)} = 0 \qquad \text{(C.31)}$$

In establishing the preceding, we make use of the following properties: (1) The limit of a ratio quantity is the limit of the quantity in the numerator to the limit of the quantity in the denominator, (2) as n increases indefinitely σ^2/n tends to zero because σ^2 is a finite number, and (3) $\Sigma x_i^2 / n \neq 0$ because the variance of X has a finite limit because of the assumption of the classical linear regression.

We leave it to the reader to show that b_1 is also a consistent estimator of B_1. (Note that $\text{var}(b_1) = \frac{\Sigma X_i^2}{n \Sigma x_i^2} \sigma^2$.)

We can extend this analysis to multiple regression: $y = XB + u$.

We have already shown that the estimator $b = (X'X)^{-1}X'y$ is an unbiased estimator of B and that variance is $\sigma^2 (X'X)$. To prove that b is a consistent estimator, we proceed as follows:

$$\text{var}(b) = \sigma^2 (X'X)^{-1}$$

$$= \frac{\sigma^2}{n} \left(n^{-1} X'X \right)^{-1}$$

Taking the probability limit of this expression as $n \to \infty$, we obtain

$$\lim_{n \to \infty} \text{var}(b) = \lim_{n \to \infty} \left[\frac{\sigma^2}{n} (n^{-1} X'X)^{-1} \right]$$

$$= \lim_{n \to \infty} \frac{\sigma^2}{n} \lim_{n \to \infty} (n^{-1} X'X)^{-1} \qquad \text{(C.32)}$$

The assumption that the elements of the matrix X are bounded implies that $n^{-1}X'X$ and hence $(n^{-1}X'X)^{-1}$ is also bounded for all n. Therefore, $\lim_{n\to\infty}(n^{-1}X'X)^{-1}$ can be replaced by a matrix of finite constants since $\lim_{n\to\infty}\dfrac{\sigma^2}{n} = 0$. This establishes the consistency of b.

C.2.2.2 Consistency of S^2:[5]

If b is a consistent estimator of B and the matrix X is nonstochastic, we can prove that S^2 is a consistent estimator of the true variance, σ^2. To show this, we proceed as follows:

$$e = y - Xb$$
$$= (XB + u) - Xb$$
$$= X(B - b) + u$$
$$(e - u) = X(B - b) \tag{C.33}$$

Now each element $(e - u)$ converges in probability to zero. That is, e_t converges in probability to u_t. As a result, the limiting behavior of $S^2 = \Sigma e_t^2 / n - k$ is equal to the limiting behavior of $\Sigma u_t^2 / n - k$, which is equal to the limiting behavior of $\Sigma u_t^2 / n$. Since the variables u_t are iid with $E(u_t^2) = \sigma^2$, then by Khinchine's theorem it follows that $p\lim(\Sigma u_t^2 / n) = \sigma^2$, which proves that S^2 is a consistent estimator of σ^2.

C.2.3 Asymptotic Efficiency

Let $\hat{\theta}$ be an estimator of θ. The variance of the asymptotic distribution of $\hat{\theta}$ is called the **asymptotic variance** of $\hat{\theta}$. If $\hat{\theta}$ is consistent and its asymptotic variance is smaller than the asymptotic variance of all other consistent estimators of θ, then $\hat{\theta}$ is called **asymptotically efficient**.

C.2.4. Asymptotic Normality

An estimator $\hat{\theta}$ is said to be asymptotically normally distributed if its sampling distribution approaches the normal distribution as the sample size n increases indefinitely. For example, statistical theory shows that if $X_1, X_2, \ldots,$ X_n are independently normally distributed with the same mean μ and the same variance σ^2, the sample mean $\bar{X}$ is also normally distributed with mean μ and

---

[5]The following discussion is based on Stewart, J., & Gill, L. (1998). *Econometrics* (2nd ed., p. 114). Upper Saddle River, NJ: Prentice Hall.

variance σ^2/n in small as well as large samples. But if values of X_i are independent with mean μ and variance σ^2 but are not necessarily from the normal distribution, then the sample mean $\bar{X}$ is *asymptotically* normally distributed with mean μ and variance σ^2/n. This in essence is one version of the **central limit theorem (CLT)**, which is of fundamental importance in statistics.

Let $X_1, X_2, \ldots, X_n$ represent a random sample from any population with $E(X) = \mu$ and var$(x) = \sigma^2$ and consider the standardized sample mean

$$Z = \frac{(\bar{X} - \mu)}{\sigma/\sqrt{n}} = \frac{\sqrt{n}(\bar{X} - \mu)}{\sigma} \tag{C.34}$$

By the CLT, Z converges in distribution to $N(0, 1)$. Equivalently, $\sqrt{n}(\bar{X} - \mu)$ converges in distribution to $N(0, \sigma^2)$.[6]

As Goldberger notes, "Approximating the cdf of the standardized sample mean Z_n by the $N(0, 1)$ cdf amounts to approximating the cdf of the sample mean $\bar{X}_n$ by the $N(\mu, \sigma^2/n)$ cdf."[7] As a result, we can write

$$\bar{X} \sim \text{asy}N(\mu, \sigma^2/n) \tag{C.35}$$

In other words, the *asymptotic distribution of* $\bar{X}_n$ is normal with mean μ and variance σ^2/n and we can call μ and σ^2/n as the *asymptotic expectation* and *asymptotic variance* of the estimator $\bar{X}_n$, where n is the sample size.

It is important to note the distinction between the *limiting* distribution of the sample mean, which is degenerate at μ (see our discussion about the consistency of an estimator), and the *asymptotic distribution* of the sample mean, which is $N(\mu, \sigma^2/n)$. As Goldberger notes, the latter provides more useful information.

In case a density function involves more than one parameter, we can express asymptotic normality in the following form:

$$\hat{\theta} \sim \text{asy}N[\theta, I^{-1}(\theta)] \tag{C.36}$$

What (C.36) says is that the estimated vector of parameters is normally distributed with mean the population parameter vector θ and the variance given by the inverse of the *information matrix* $I(\theta)$, which is defined as

---

[6]For proof, see DeGroot, M. H. (1975). *Probability and statistics.* Reading, MA: Addison-Wesley. The proof involves moment-generating or characteristic functions. It may be added that there are several versions of the CLT.

[7]Goldberger, A. S. (1991). *A course in econometrics* (p. 99). Cambridge, MA: Harvard University Press.

$$I(\theta) = E\left[\left(\frac{\partial l}{\partial \theta}\right)\left(\frac{\partial l}{\partial \theta}\right)'\right] = -E\left[\frac{\partial^2 l}{\partial \theta \partial \theta'}\right] \tag{C.37}$$

where l is the log-likelihood function.

If θ is a vector of k elements, $\frac{\partial l}{\partial \theta}$ is a column vector of k partial derivatives, that is,

$$\frac{\partial lf}{\partial \theta} = \begin{bmatrix} \partial lf / \partial \theta_1 \\ \partial lf / \partial \theta_2 \\ \vdots \\ \partial lf / \partial \theta_k \end{bmatrix} \tag{C.38}$$

Each element in this **score** or **gradient vector** is itself a function of θ and may be differentiated partially with respect to each element in θ. As an example,

$$\frac{\partial}{\partial \theta_1}(\partial l / \partial \theta) = \begin{bmatrix} \dfrac{\partial^2 l}{\partial \theta_1^2} \\ \dfrac{\partial^2 l}{\partial \theta_1 \partial \theta_2} \\ \vdots \\ \dfrac{\partial^2 l}{\partial \theta_1 \partial \theta_k} \end{bmatrix} \tag{C.39}$$

Proceeding in this manner, we obtain the following square symmetric matrix of second-order derivatives, which is known as the **Hessian matrix** that we encountered earlier.

$$\frac{\partial^2 l}{\partial \theta \partial \theta'} = \begin{bmatrix} \dfrac{\partial^2 l}{\partial^2 \theta_1^2} & \dfrac{\partial^2 l}{\partial \theta_1 \partial \theta_2} & \cdots & \dfrac{\partial^2 l}{\partial \theta_1 \partial \theta_k} \\ \dfrac{\partial^2 l}{\partial \theta_2 \partial \theta_1} & \dfrac{\partial^2 l}{\partial \theta_2^2} & \cdots & \dfrac{\partial^2 l}{\partial \theta_2 \partial \theta_k} \\ \vdots & \ddots & \ddots & \vdots \\ \dfrac{\partial^2 l}{\partial \theta_k \partial \theta_1} & \dfrac{\partial^2 l}{\partial \theta_k \partial \theta_2} & \cdots & \dfrac{\partial^2 l}{\partial \theta_k^2} \end{bmatrix} \tag{C.40}$$

APPENDIX D: SOME IMPORTANT PROBABILITY DISTRIBUTIONS[1]

Estimation and hypothesis testing are two important branches of *statistical inference*. We have discussed estimation of the linear regression model using ordinary least squares and maximum likelihood methods of estimation. Hypothesis testing requires developing appropriate **test statistics** and their probability distributions that will aid us in testing hypotheses and establishing confidence intervals for the parameters of interest. We now discuss several probability distributions and their properties that will help us in developing appropriate test statistics.[2]

D.1 The Normal Distribution and the Z Test

Let $X_i \sim N(\mu, \sigma^2)$, that is, a normally distributed random variable with mean μ and variance σ^2. Its probability density function is

$$f(X) = \frac{1}{\sigma\sqrt{2\pi}} \exp^{-\frac{1}{2}\left(\frac{X-\mu}{\sigma}\right)^2} \text{ where } \sigma > 0 \tag{D.1}$$

Now consider the following variable:

$$Z_i = \frac{X_i - \mu}{\sigma} \tag{D.2}$$

Z_i is a **standardized normal variable**, which has zero mean and unit variance, that is,

$$Z_i \sim N(0,1) \tag{D.3}$$

---

[1]For derivations of the distributions discussed in this appendix, see any textbook in mathematical statistics. For example, see Hogg, R. V., Mckean, J., & Craig, A. T. (2012). *Introduction to mathematical statistics* (7th ed.). Harlow, England: Pearson Education.

[2]For an application of various statistical tests, see Kanji, G. K. (1999). *100 statistical tests* (New ed.). London, England: Sage.

220

D.1.1 Properties of the Normal Distribution

1. It is symmetric around its mean value.

2. Approximately 68% of the area under the normal curve lies between the values of $\mu \pm \sigma$, about 95% of the area lies between $\mu \pm 2\sigma$, and about 99.7% of the area lies between $\mu \pm 3\sigma$ (see Figure D.1).

3. It depends on only two parameters, mean μ and variance σ^2.

4. If $X_1 \sim N(\mu_1, \sigma_1^2)$ and $X_2 \sim N(\mu_2, \sigma_2^2)$, and they are independent, then

$$Y = (aX_1 + bX_2) \sim N[(a\mu_1 + b\mu_2), (a^2\sigma_1^2 + b^2\sigma_2^2)] \qquad \text{(D.4)}$$

In words, *a linear combination of normally distributed variables is also normally distributed.* This result can be generalized to linear combinations of more than two independently distributed normal variables.

5. The third and fourth moments of the normal distribution around the mean value are as follows:

$$\text{Third moment: } E(X - \mu)^3 = 0$$

$$\text{Fourth moment: } E(X - \mu)^4 = 3\sigma^4$$

Note: All odd-powered moments about the mean value of a normally distributed variable are zero.

Figure D.1   The Normal Distribution

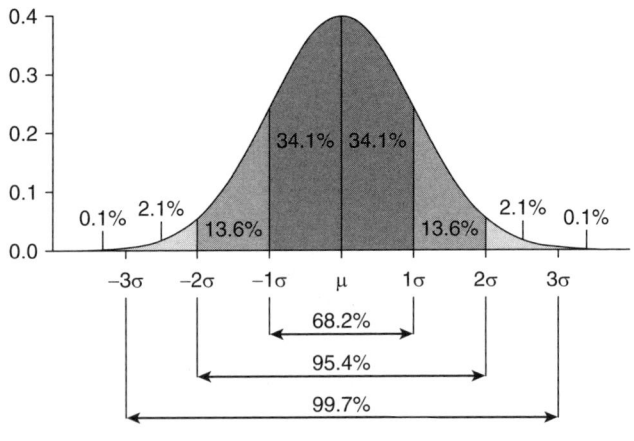

6. As a result, for a normally distributed variable, the **skewness coefficient** (S) is zero and the **kurtosis coefficient** (K) is 3. Skewness is a measure of asymmetry of a probability distribution, and kurtosis is a measure of how tall or flat the probability distribution is.

7. A simple test of normality is to find out whether the skewness coefficient and kurtosis measures are 0 and 3, respectively. This is the basis of the **Jarque–Bera (JB) test of normality**, which is defined as follows:

$$JB = n\left[\frac{S^2}{6} + \frac{(K-3)^2}{24}\right] \sim \chi_2^2 \qquad (D.5)$$

Under the null hypothesis of normality, the JB statistic is distributed as a chi-square statistic with 2 *df*. Notice that the JB statistic is a test of the *joint hypothesis* that the skewness coefficient is zero and the kurtosis coefficient is 3. That is the reason for the 2 *df*.

8. The mean and variance of a normally distributed random variable are independent, that is, one is not a function of the other.

9. If X and Y are jointly normally distributed, then they are independent, if and only if the covariance between them is zero.

10. **The central limit theorem (CLT)**: Let $X_1, X_2, \ldots, X_n$ denote n independent random variables, all of which have the same probability distribution function with mean = μ and variance = σ^2. Let $\bar{X} = \Sigma X_i / n$ be the sample mean. Then as $n \to \infty$,

$$\bar{X} \sim N(\mu, \sigma^2 / n) \qquad (D.6)$$

That is, $\bar{X}$ approaches the normal distribution with mean μ and variance σ^2 / n. This result holds true regardless of the form of the probability distribution function. As a result, it follows that

$$Z = \frac{\bar{X} - \mu}{\sigma / \sqrt{n}} = \frac{\sqrt{n}(\bar{X} - \mu)}{\sigma} \sim N(0,1) \qquad (D.7)$$

That is, Z is a standardized normal variable.

D.2 The Gamma Distribution

A random variable X has a gamma distribution if its probability distribution function is as follows:

$$f(X) = \frac{1}{\beta^\alpha \Gamma(\alpha)} X^{\alpha-1} \exp(-X / \beta) \text{ for } X > 0$$

$$= 0 \text{ elsewhere}$$

(D.8)

where $\alpha > 0$ and $\beta > 0$ (see Figure D.2).

The gamma function, $\Gamma(\alpha)$, is defined as

$$\Gamma(\alpha) = \int_0^\infty Y^{\alpha-1} e^{-Y} dy \quad \text{for } \alpha > 0$$

(D.9)

Using calculus techniques (integration by parts), it can be shown that the gamma function satisfies the recursive formula

$$\Gamma(\alpha) = (\alpha - 1)\Gamma(\alpha - 1) \quad \text{for } \alpha > 1$$

(D.10)

Figure D.2   The Gamma Distribution

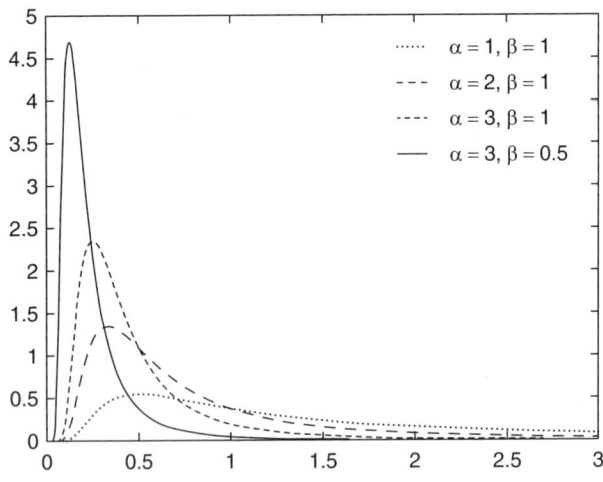

Note that

$$\Gamma(1)=\int_0^\infty e^{-Y}\mathrm{d}y=1 \tag{D.11}$$

Then, by repeated application of the recursive formula it follows that

$$\Gamma(\alpha)=(\alpha-1)! \tag{D.12}$$

where α is a positive integer.

An important special case is

$$\Gamma(\frac{1}{2})=\sqrt{\pi} \tag{D.13}$$

Special cases of the gamma distribution are the **chi-square**, **Erlang**, and **exponential distributions**.

D.3 The Chi-Square (χ^2) Distribution and the χ^2 Test

Let $Z_1,Z_2,\ldots,Z_k$ be independent standardized normal variables [i.e., $Z_i \sim N(0,1)$].

Then, the quantity

$$Z=\Sigma Z_i^2 \sim \chi_k^2 \tag{D.14}$$

is said to possess the χ^2 distribution with k df, where the df means the number of independent quantities in the previous sum.

A test statistic based on the chi-square distribution is called a **chi-square test**.

The probability distribution function of the chi-square distribution is as follows:

$$f(X)=\frac{X^{(k-2)/2}\exp(-X/2)}{2^{k/2}\Gamma(k/2)} \quad \text{for } X>0$$
$$=0\,\text{elsewhere} \tag{D.15}$$

where $\Gamma(k/2)$ is the **gamma function** with argument $k/2$, where k represents the df (see Figure D.3).

Figure D.3   The Chi-Square Distribution for Select Degrees of Freedom

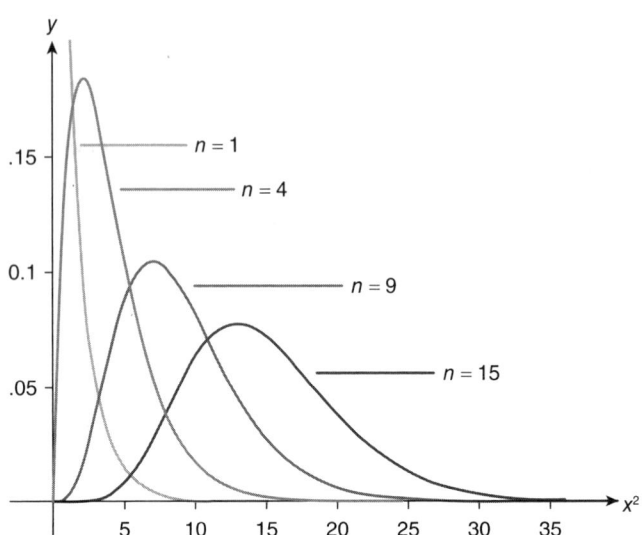

D.3.1 Properties of the Chi-Square Distribution

1. The range of X is $0 \le X < \infty$.
2. It is a skewed distribution, the degree of the skewness depending on the *df*.
3. Its mean value is k and its variance is $2k$, a unique property of this distribution.
4. Its mode is $k - 2$, $k > 2$.
5. Its median is $k - 2/3$ (approximately for large k).
6. Its coefficient of skewness is $2^{3/2} k^{-1/2}$.
7. Its coefficient of kurtosis is $3 + 12/k$.
8. As the *df* increases, the chi-square distribution becomes increasingly symmetrical. As a matter of fact, for *df* in excess of 100, the variable

$$\sqrt{2\chi^2} - \sqrt{(2k-1)} \sim N(0,1) \tag{D.16}$$

can be treated as a standardized normal variable, where k is the *df*.

9. If Z_1 and Z_2 are two independent chi-square variables, with k_1 and k_2 *df*, respectively, then the sum $Z_1 + Z_2$ is also a chi-square variable with $df = k_1 + k_2$.

D.4 Student's *t* Distribution

If Z_1 is a standardized normal variable [i.e., $Z_1 \sim N(0,1)$] and another variable Z_2 follows the chi-square distribution with k df and is independent of Z_1, then

$$t = \frac{Z_1}{\sqrt{Z_2 / k}}$$

$$= \frac{Z_1 \sqrt{k}}{\sqrt{Z_2}} \sim t_k \tag{D.17}$$

A test based on the *t* test is called (Student's) **t test**.

The properties of the *t* distribution are as follows:

1. The *t* distribution, like the normal distribution, is symmetrical, but it is flatter than the normal distribution. But as the *df* increases, the *t* distribution approximates the normal distribution. The approximation is reasonable for $k > 30$.

2. The mean of the *t* distribution is zero, and its variance is $k/(k - 2)$, which exists if $k > 2$.

3. Coefficient of skewness = 0, $k > 3$.

4. Coefficient of kurtosis = $3(k - 2)/(k - 4)$, $k > 4$.

5. If $X_1, X_2, \ldots, X$ are iid $N(\mu, \sigma^2)$ random variables, then

$$\frac{\bar{X} - \mu}{s} \sqrt{n} \sim t_{n-1} \tag{D.18}$$

where S^2 = sample variance.

Figure D.4 The Normal and *t* Distributions

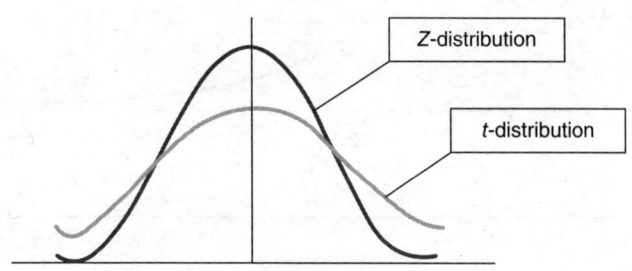

Z-distribution

t-distribution

6. The t distribution with 1 df is called the Cauchy distribution (see Figure D.4).

D.5 Fisher's F Distribution

If Z_1 and Z_2 are independently distributed chi-square variables with k_1 and k_2 df, respectively, the variable

$$F = \frac{Z_1 / k_1}{Z_2 / k_2} \sim F_{k_1 k_2} \qquad (D.19)$$

follows the F distribution (see Figure D.5).

Note: k_1 is called the *numerator df* and k_2 is called the *denominator df*.

A test based on the F distribution is called an **F test**.

D.5.1 Properties of the F Distribution

1. Like the chi-square distribution, the F distribution is skewed to the right. But it can be shown that as k_1 and k_2 become increasingly larger, the F distribution approaches the normal distribution.

Figure D.5   The F Distribution for 4 and 10 Degrees of Freedom

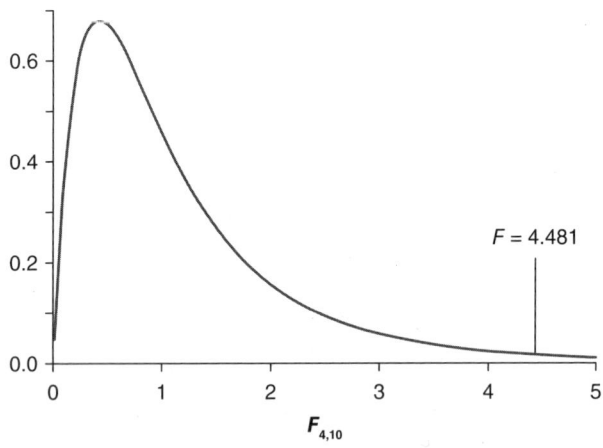

2. The mean value of an F-distributed variable is $k_2/(k_2 - 2)$, which is defined for $k_2 > 2$, and its variance is

$$\frac{2k_2^2(k_1 + k_2 - 2)}{k_1(k_2 - 2)^2(k_2 - 4)} \tag{D.20}$$

which is defined for $k_2 > 4$.

3. Coefficient of skewness is

$$\frac{(2k_1 + k_2 - 2)[8(k_2 - 4)]^{1/2}}{k_1^{1/2}(k_2 - 6)(k_1 + k_2 - 2)^{1/2}} \quad k_2 > 6 \tag{D.21}$$

4. Coefficient of kurtosis is

$$3 + \frac{12[(k_2 - 2)^2(k_2 - 4) + k_1(k_1 + k_2 - 2)(5k_2 - 22)]}{k_1(k_2 - 6)(k_2 - 8)(k_1 + k_2 - 2)} \quad \text{for } k_2 > 8 \tag{D.22}$$

5. The square of a t-distributed random variable with k df has an F distribution with 1 and k df. That is,

$$t_k^2 = F_{1,k} \tag{D.23}$$

6. If the denominator df k_2 is fairly large, then we have the following relationship:

$$k_1 F \sim \chi_{k1}^2 \tag{D.24}$$

In words, for a fairly large denominator df, the numerator df times the F value is approximately the same as a chi-square value with the numerator df.

7. If s_1^2 and s_2^2 are the variances of independent random samples of size n_1 and n_2 from normal populations with variances σ_1^2 and σ_2^2, then

$$F = \frac{s_1^2/\sigma_1^2}{s_2^2/\sigma_2^2} = \frac{\sigma_2^2 s_1^2}{\sigma_1^2 s_2^2} \sim F_{n_1-1,n_2-1} \tag{D.25}$$

D.6 Relationships Among Probability Distributions

$$F_{1,k} = t_k^2 \qquad \text{(D.26)}$$

That is, the square of the t statistic with k df is equal to an F statistic with 1 df in the numerator and k df in the denominator.

$$mF_{m,n} = \chi_m^2 \quad n \to \infty \qquad \text{(D.27)}$$

That is, for large denominator df, the numerator df times the F value is approximately equal to the chi-square value with the numerator df, where m is the numerator df and n is the denominator df.

$$Z = \sqrt{2\chi^2} - \sqrt{2k-1} \sim N(0,1) \qquad \text{(D.28)}$$

This states that for sufficiently large df, the chi-square distribution can be approximated by the standard normal distribution, where k is df.

D.7 Uniform Distributions

There are two types of uniform distributions: (1) discrete uniform and (2) continuous uniform.

D.7.1 Discrete Uniform Distribution

Let $n > 1$ be an integer. Suppose the variable X has mass function given by

$$f(X) = 1/n \quad \text{for } X = 1, 2, \ldots, n$$
$$= 0 \text{ otherwise} \qquad \text{(D.29)}$$

We say that X has a uniform distribution on $(1, 2, \ldots, n)$ (see Figure D.6).

$$\text{Mean of } X = (n+1)/2 \qquad \text{(D.30)}$$
$$\text{Variance of } X = (n^{2-1})/12 \qquad \text{(D.31)}$$

D.7.2 Continuous Uniform Distribution

The continuous random variable X has a uniform distribution over the interval (a, b) if its density function is given by

$$f(X) = \begin{cases} \dfrac{1}{b-a} & a < x < b \end{cases}$$
$$= 0 \text{ otherwise} \qquad \text{(D.32)}$$

Figure D.6 Discrete Uniform Distribution

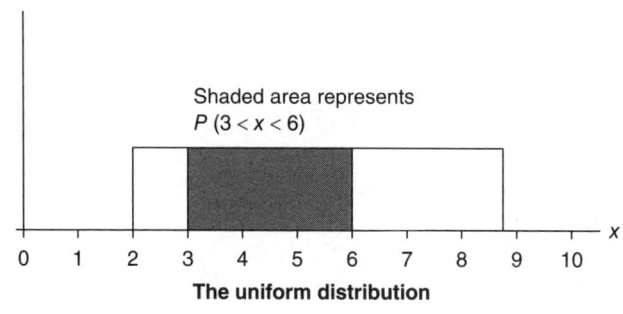

The uniform distribution

$$F(X) = \begin{cases} 0 & \text{for } x < a \\ \dfrac{x-a}{b-a} & \text{for } a \le x < b \\ 1 & \text{for } x \ge b \end{cases}$$

$$\text{mean} = (a+b)/2 \tag{D.33}$$

$$\text{variance} = (b-a)^2 / 12 \tag{D.34}$$

D.8 Some Special Features of the Normal Distribution[3]

If $X_1, X_2, \ldots, X_n$ is a sample from the normal population with mean $= \mu$ and variance $= \sigma^2$, that is, $X_i \sim N(\mu, \sigma^2)$, then the following features apply:

1. The sample mean $\bar{X}$ and the sample variance S^2 are independent random variables.

2. $\bar{X} \sim N(\mu, \sigma^2 / n)$.

3. $(n-1)\dfrac{S^2}{\sigma^2} \sim \chi^2_{n-1}$.

---

[3]The proofs of some of the following statements are rather involved. See Casella, G., & Berger, R. L. (2000). *Statistical inference* (2nd ed.) Belmont, CA: Wadsworth; Rice, J. A. (2007). *Mathematical statistics and data analysis* (3rd ed.). Pacific Grove, CA: Brooks/Cole; Wasserman, L. (2005). *All of statistics: A concise course in statistical inference*. New York, NY: Springer Science & Business Media.

The proof of Statement 1 is slightly complicated (see references in Footnote 1). So for now, we will assume that this is the case.

Statement 2 is easy to establish:

$$E(\overline{X}) = \frac{1}{n} E\left(\sum_1^n X_i \right) = \frac{1}{n}(n \cdot \mu) = \mu \tag{D.35}$$

since the expectation of each X_i is μ.

$$\begin{aligned} \operatorname{var} \overline{X} &= \frac{1}{n^2} \operatorname{var}(X_1, X_2, \ldots, X_n) \\ &= \frac{1}{n^2}(n \cdot \sigma^2) = \frac{\sigma^2}{n} \end{aligned} \tag{D.36}$$

since the Xs are independently distributed, each with variance of σ^2.

To establish Statement 3 above, we start with the following identity:

$$\sum_{i=1}^n (X_i - \overline{X})^2 = \sum_{i=1}^n (X_i - \mu)^2 - n(\overline{X} - \mu)^2 \tag{D.37}$$

which we can express as

$$\frac{\sum_{i=1}^n (X_i - \mu)^2}{\sigma^2} = \frac{\sum_{i=1}^n (X_i - \overline{X})^2}{\sigma^2} + \frac{n(\overline{X} - \mu)^2}{\sigma^2} \tag{D.38}$$

which is equivalent to

$$\sum_{i=1}^n \left(\frac{X_i - \mu}{\sigma} \right)^2 = \frac{(n-1)S^2}{\sigma^2} + \left[\frac{\sqrt{n}(\overline{X} - \mu)}{\sigma} \right]^2 \tag{D.39}$$

Noting that $\left(\dfrac{X_i - \mu}{\sigma} \right)$ is a unit normal variable so its square is a chi-square random variable with 1 df and since the left-hand side of Equation (D.39) is the sum of n independent chi-square random variables, it has n df. The last term in Equation (D.39) is also a chi-square variable with 1 df.

Now since $\overline{X}$ and S^2 are independent by Statement 1 above, it follows the two terms on the right-hand side of Equation (D.39) are independent, hence the conclusion that $\dfrac{(n-1)S^2}{\sigma^2}$ has a chi-square distribution with

$(n-1)$ $df.$[4] Recall that the sum of two or more independent chi-square variables is also the chi-square variable with degrees of freedom equal to the sum of the degrees of freedom of associated chi-square variables.

4. $\dfrac{\bar{X} - \mu}{s / \sqrt{n}} \sim t_{(n-1)}$ $\hspace{6cm}$ (D.40)

5. If $X_i \sim N(\mu_x, \sigma_x^2)$ and $Y_i \sim N(\mu_y, \sigma_y^2)$ and X_i and Y_i are independent of each other, then

$$\dfrac{S_x^2 / S_y^2}{\sigma_x^2 / \sigma_y^2} \sim F_{(n,m)} \hspace{5cm} \text{(D.41)}$$

where S_x^2 and S_y^2 are the sample variances of X and Y and n and m are their associated degrees of freedom.

The proofs of Equations (D.40) and (D.41) can be found in the references cited in Footnotes 1 and 4.

---

[4]For a rigorous proof of this statement, see Casella, G., & Berger, R. L. (2002). *Statistical inference* (2nd ed., pp. 218–219). Belmont, CA: Wadsworth.

INDEX